Adrie Winkelaar

Coatings Basics

Adrie Winkelaar: Coatings Basics
© Copyright 2009 by Vincentz Network, Hannover, Germany
ISBN 978-3-86630-851-0

Cover: Evonik Degussa Coatings & Colorants –
Dieter Debo, Infracor, Marl/Germany

Adrie Winkelaar
Coatings Basics
Hannover: Vincentz Network, 2009
(European Coatings Tech Files)
ISBN 3-86630-851-5
ISBN 978-3-86630-851-0

Please ask for our book catalogue
Vincentz Network, Plathnerstr. 4c, 30175 Hannover, Germany
T (202) 684-6630, F (202) 380-9129
E-mail: books@american-coatings.com, www.american-coatings.com

Layout: Maxbauer & Maxbauer, Hannover, Germany

ISBN 3-86630-851-5
ISBN 978-3-86630-851-0

European Coatings Tech Files

Adrie Winkelaar

Coatings Basics

Adrie Winkelaar: Coatings Basics
© Copyright 2009 by Vincentz Network, Hannover, Germany
ISBN 978-3-86630-851-0

Preface

The purpose of this book is to provide an insight to the development, manufacture and application of paint products for those who have little or no education in coating technologies. Paint is considered to be a straightforward product and simple to apply, however, all paints contain a variety of risk-involved chemicals that are in accordance with physical and chemical laws. Increasingly more safety, health and environment legislations have been and are being passed in order to protect paint users, such as professional painters, do-it-yourselfers and industrial applicators.

The physical and chemical processes when applying paint products are extremely important, it is therefore necessary for the user to understand what the properties are and what to expect so that the best possible protection is provided and disappointments of substrate decoration results are avoided. This book, comprising ten chapters, offers important and useful explanations and guidelines to understanding what happens and what to expect when paints are developed, manufactured and applied.

The following chapters have been divided into various topics, starting in Chapter 1 with a short summary of the lengthy history of paint, worldwide differences and definitions. Because paint products contain chemicals that are in accordance to physical and chemical laws some basic chemistry information is necessary for understanding the background and basics of paint. The book continues with information on ingredients and properties of paint. It is important to know something about the various paint formulas and the differences between wall paint, wood paint, and metal paint etc. Paint manufacture and paint application are steps in the process of understanding how sensitive paint can react. The final two chapters of this book are concerned with test methods and regulations for health, safety and

Adrie Winkelaar: Coatings Basics
© Copyright 2009 by Vincentz Network, Hannover, Germany
ISBN 978-3-86630-851-0

environment. Solvents in paints are harmful to health, safety and to the environment and are gradually being reduced and replaced with water. Increasingly new technologies are being introduced in the paint industry. Hopefully this book will provide an insight, not only to the present properties and use of paint but also to future new developments.

This book has been written with very few references so that newcomers are able to read it through freely and easily. Many illustrations and figures have been inserted to show the various manufacturing machinery and test methods equipment. The appendix contains a list of reference books that provide more specific, extensive knowledge about organic chemical science. It is always possible to provide more detailed information regarding ingredients, formulations, manufacturing, application and legislation, however, this book offers concise information and guidelines for understanding the paint and coatings phenomena.

Adrie Winkelaar
Heemskerk, Netherlands, May 2009

Contents

Adrie Winkelaar: Coatings Basics
© Copyright 2009 by Vincentz Network, Hannover, Germany
ISBN 978-3-86630-851-0

1 What is paint or coating?

To explain the many varieties of paint applications that are world-wide available, it is important to understand the rich and extensive history of paint. Paint is a personal experience to which emotions and feelings are linked and which vary worldwide. Architectural paint provides a new fresh appearance to many used things and because the question of what is beautiful is different for each and every one of us, it is a personal issue. Looking back through the ages we have learned that throughout the world of paint has varied. It is now possible to see how paint differs from country to country and perhaps we are gradually moving towards a common worldwide feeling towards paint by using the same compositions.

The first history questions are: When was the first paint made? How was the first composition devised? What was the first application?

Through ages paint has been more and more developed with the use of new experience which resulted in making new applications possible. The increasing introduction of different pigments, different dyes and different binders has resulted in an increase of paint applications that have provided even more possibilities to communicate worldwide. It is very important to communicate durability and sustainability to all cultures and future generations throughout the world.

Present-day coatings appear on all objects in our environment. Knowing something about paint makes for being an authority on our 'painted = coated' surroundings. Coatings are applied to houses, cars, aeroplanes, furniture, and computers – paint is, in fact, applied to so many objects and in so many different ways. There are also many similarities between paint and cosmetics, paint and wallpaper, paint and glazing and between paint and foil. Basic knowledge about these substances will open up a new, wonderful and colourful world. Colour is coating and coating is colour.

Adrie Winkelaar: Coatings Basics
© Copyright 2009 by Vincentz Network, Hannover, Germany
ISBN 978-3-86630-851-0

Differences between paint and coating

When the substance is wet – it is paint and when it is a dried layer – it is a coating. People sometimes call thick layers "coatings" and thin layers "paint", such as paint on wood. Wet paint products for thick layers (for metal protection) are also called "coatings". Generally speaking – wet paints are in a tin or layer and "coatings" are dried layers on substrates.

1.1   History

Paint goes far back in history. According to neuropsychologists, human beings developed their language between 50,000 and 70,000 years ago and the first paintings were discovered after that period. A language had been developed to enable communication and paint is in that case the first visual expression of communication. It is estimated that the first paint was composed between 30,000 to 40,000 years ago. The oldest cultures used carbonized wood to illustrate expressions of what they saw, such as animals and landscapes. The Cro-Magnon men painted on the walls in caves in France (Lascaux) and Spain (Altamira) between 15,000 and 10,000 B.C. Ferrous earth was also used as pigment and blood and milk as binders.

Throughout Europe many cave paintings have been found – from Spain and right across to The Ural. The oldest painting which was found in the Lions Cave, South Africa dates back to 40,000 years ago.

Figure 1.1: Cave paintings in Lascaux

Paint was later used to identify and decorate objects, clothes and the body. The natural substances used are still being applied for re-usable binding substances in the modern paint industry, such as casein, grease, waxes and resins from milk, plants and trees.

The advanced civilisation of the Egyptians, Chinese, Greeks and Romans used painting techniques to identify and decorate vessels, statues, tools and buildings. Paint was also increasingly being used to provide protection against the influences of weather. Wooden ships were made watertight with mixtures of natural bitumen and asphalt. The old Greeks used the first anti-fouling application to protect their ships against under water fouling, which increased the speed of their vessels compared to enemy ships.

Figure 1.2: Vessels in original colours from the Ancient Greece period

Approximately 2000 years before Christ marks an important period in the development of paint which is illustrated by the decoration techniques of the Ancient Chinese who produced smooth and glossy objects. Varnishes offered a new dimension to paintwork and the new raw materials such as balsams and natural resins provided many possibilities. The most famous resin is shellac, produced by certain insects called 'lac insects' that produce sap in Indian fig trees.

In the 11th century, monks made a spectacular step forward in the history of paint when they boiled linseed oil with molten amber and acquired a more durable coating. From that period coatings were used to maintain paintings, the shields of knights and as durable protection of wood, in addition to the old fashion bitumen and asphalt. This resulted in a unity of decoration and protection and thus the manufacture of paint as a trade was born. The famous painters in the 15th, 16th and 17th centuries made their own paints. At that time each part of the world had its own local artist who made paint and applied the colours to the inside and outside of houses and public buildings.

In the 18th century the industrial revolution caused a huge demand for paint for all objects, houses and ships. The world population increased and many cities and villages evolved. At that time the first paint factories were opened in England, Holland and Germany. At these factories investigations were carried out using brewing kettles for the manufacturing of linseed oil binders and machines for milling the pigments.

The final mega development came in the 20th century, when the industry produced new such as

- nitrocellulose,
- alkyds,
- acrylics,
- polyurethanes and
- epoxies

as binders for use in the paint industry. New production lines were opened for the manufacture of numerous consumer goods such as cars, buses and trains, but also for furniture, beds and paint for the decorative market. For all these applications the paint industry used micro-technology with of 1 to 10 micron. One micron is a one thousand part of a millimetre. At the beginning of the 21st century new technology introduced nano-particles consisting of new properties. A nanometre is a one thousand part of a micron. It is possible to introduce nano-particles into micro-particles for 'self healing' paints. The nano-particles open when a scratch occurs. "Smart coatings" have also been developed as structured coating systems that provide an optimum response to certain external stimuli, and react to outside conditions, such as temperature, stress, strain or the environment, in selective ways.

The production and the use of paint has developed together with humans and dates back from prehistoric times when it was used as an experimental communication tool, through the Middle Ages when it became a trade that introduced durable properties and right up to Modern Times resulting in an ever growing multi-disciplinary high coating technology.

1.2   Worldwide differences

Paint is used on a large scale throughout the world. The paint quantity per population head varies for each country. Worldwide – Ger-

Table 1.1: European paint production and consumption in 1996, source: CEPE, the European Paint Makers Association in Brussels, www.cepe.org

	Inhabitant in millions	Paints in 1000 tonnes	kg/ inhabitant
Germany	80.4	1,350	16.8
Spain	40.4	630	15.6
France	63.7	690	10.8
Italy	58.1	730	12.6
United Kingdom	60.7	410	6.7
The Netherlands	15.5	310	20.0
Denmark	5.4	130	24.0
Belgium	10.4	140	13.5

Table 1.2: European Decorative market in 2006

	kt deco	kg/ inhabitant
Germany	800	9.7
Spain	700	17.3
France	500	7.8
Italy	400	6.9
United Kingdom	400	6.5
The Netherlands	150	9.1
Denmark	70	13.0
Belgium	50	4.8

many, Denmark and The Netherlands use the highest quantities with approximately 20 kg per inhabitant, see Table 1.1. For example: Egypt = 3 kg per inhabitant and Russia = 4,5 kg per inhabitant. The quantities of paint applications used provide an indication of the prosperity and hygiene of a certain country. The figures also include consumption and provide a total impression of welfare and economics.

In Europe the decorative market also varies for each country. This market offers an impression of habitation and public buildings in each country. The paint quantity per population head provides

Table 1.3: Decorative consumption in tonnes in 2004 in The Netherlands, source: Dutch Paint Makers Association in The Hague, The Netherlands, www.vvvf.nl

	Professional painters	Do-it-yourself	Total
Lacquers and varnishes	13,700	13,100	26,800
Wall paints	22,900	40,800	63,700
Plaster	48,500	12,700	61,200
Wood stains		1,300	1,300
Concrete repair	2,200		2,200
Fillers	2,000	850	2,850
Total	89,300	68,750	158,050

another impression. Spain and Denmark show the highest figures and provide an impression of building and maintenance activities at a specific moment, see Table 1.2.

The different habits and cultures of each country are also illustrated by the figures of the various paint types sold on the decorative market. Wall paints are applied more in Germany and in southern countries than in other countries. In northern countries a lot of wood paint and wood stains are applied.

In The Netherlands many different products, such as wood stains, lacquers, varnishes, wall paints for ceilings, interior walls, kitchens and exterior walls, etc are applied. In this country there is a difference between the professional decorative market and the do-it-yourself decorative market, see Table 1.3. The decorative market is 80 % of the total volume in The Netherlands.

The European paint makers association, CEPE, represents 85% of the European paint volume. CEPE represents 900 members; however there are approximately 3,300 paint manufactures in Europe. In the European paint industry in 2006, 120,000 people worked together and produced more than 4,000 kt of products with a total value of 17 billion Euros. The European decorative paint market is approximately 60 % of the total volume and represents about 7 billion Euros divided under 2.5 million professional painters and about 100 million do-it-yourselves. The paint impact on daily life is enormous and a world without paint is unimaginable.

1.3   *Definitions and standardization*

A coating is a dried paint on a substrate. Our environment is full of coatings – it is the most 'looked-at' product in the world. Paint definitions are described in EN/ISO 4618 and paints are applied as coatings on substrates for identification, decoration and protection purposes.

In addition, an important property of a coating is indicated with information using colours in the design and the printing trades.

There is an increasing global harmonization regarding definitions, test methods and regulations. The International Organization of Standardization (ISO) is the world's largest developer of international standards. ISO is a network of the national standards institutes of 160 countries, one member per country, with a Central Office in Geneva, Switzerland, that coordinates the systems. ISO is a non-governmental organization that forms a bridge between the public and private sectors. The ISO 9000 is a standard quality system and the ISO 14000 is a standard environment system.

In the middle of the 1990's the paint industry introduced a worldwide responsible care system based on the ISO system, called Coatings Care®. The Coatings Care® program was conceived as a voluntary initiative aimed at assisting industry professionals in their efforts to protect the health and environment of the worker and the community, as well as promote product safety. Coatings Care® is designed to foster best management practices, and promote the development of new technologies that improve product performance. The framework and resources for the program deliver comprehensive guidance for manufacturing operations as well as critical support for customers and business partners throughout the supply and distribution chain. Coatings Care® is tailor-made to be the most effective and practical system for the coatings industry to sustain safe and environmentally friendly operations around the globe.

2   *Basic principles of chemistry*

Chemistry originated in Ancient Egypt. The name chemistry derives from the Egyptian word "kēme" (chem), meaning "earth". Chemistry is the science concerned, not only with the composition, structure and properties of substances, but also the changes that substances undergo during chemical reactions. Modern chemistry results from alchemy practiced during the chemical revolution in the 18th and 19th century.

2.1   *Substances*

The concept of a chemical element is related to that of a chemical. A chemical element is characterized by its atom – the basic unit of an element. A collection of matter consisting of a positively charged core (the atomic nucleus) made up of protons and neutrons. A number of electrons, surrounding the atomic nucleus, are maintained in order to balance the positive charge within the nucleus. An electron has a negative charge. An atom is also the smallest entity that can be envisaged and retains some of the chemical properties of the element, such as ionization potential and a preferred oxidation state. Each element has a characterized element value with a number of protons.

A presentation of the chemical elements is illustrated in the "periodic table" (see Appendix 1), which groups elements by atomic values. *Dmitri Mendeleev* (1869) was the first scientist to create a periodic table of elements. This table shows that when elements were ordered by an increasing atom weight (protons and neutrons), a pattern appeared when the properties of the elements were periodically repeated. There are more than 103 elements, many of which are important to daily life, such as, number 1 (H = Hydrogen), number 6 (C = Carbon), number 8 (O = Oxygen), number 26 (Fe = Iron), number 47 (Ag = Silver) and number 79 (Au = Gold).

Adrie Winkelaar: Coatings Basics
© Copyright 2009 by Vincentz Network, Hannover, Germany
ISBN 978-3-86630-851-0

Apart from an atom, a molecule is the smallest indivisible portion of a pure chemical substance that has a unique collection of chemical properties. Molecules are a typical set of atoms bound together by covalent bonds, such that, the structure is electrically neutral and all of the valence electrons are paired with other electrons either in bonds or in lone pairs.

A molecule may consist of atoms of the same chemical element, such as with oxygen (O_2), or of different elements, such as with water (H_2O). Atoms and complexes connected by non-covalent bonds such as hydrogen bonds or ionic bonds are not generally considered as single molecules.

One of the main characteristics of a molecule is its geometry – often called its 'structure'. No typical molecule can be defined in ionic crystals (salts) and covalent crystals (network solids), although these are often composed of repetitive unit cells that extend either in a plane (such as in graphite) or three-dimensionally (such as in diamonds or sodium chloride). The theme of the repetitive unit-cellular-structure also applies to the most condensed phases of metallic bonding. In glass (solids that exist in a vitreous disordered state), atoms may also

Table 2.1: Chemical description of some substances

Structure	Angle of atoms	Formula	Visualisation
Linear	180°	beryllium-chlorine: $BeCl_2$	
Trigonal planar	120°	boron-fluorine: BF_3	
Bent	120°	sulphur dioxide: SO_2	
Tetrahedral	109.5°	methane: CH_4	
Trigonal pyramidal	109.5°	ammonia: NH_3	
Bent	109.5°	water: H_2O	

be held together, not only by chemical bonds without any definable molecule, but also without any of the regularity of repetitive units that characterize crystals.

In each molecule structure the elements are bonded with other elements in a specific geometry, such as linear, pyramidal or rings. The chemical formula provides the letters of the elements and each formula can be visualised.

In the following Table 2.1 the chemical structures of some substances are described by name, formula, and angles of the atoms and the visualisation of the structure.

2.2   Organic (carbon) chemistry

Carbon is found in many different compounds. It is in the food we eat, the clothes we wear, the cosmetics we use and the gasoline that fuels our car. Carbon is the sixth most common element in the universe. In addition, carbon is a very special element because it plays a dominant role in the chemistry of life.

The simplest organic compounds contain molecules composed of carbon and hydrogen. The compound methane contains one carbon bonded to four hydrogens. Ethane is another example of a simple hydrocarbon. Ethane contains two carbon atoms and six hydrogen atoms. In chemistry we use a molecular formula to show how many atoms of each element are present in a molecule. However, a molecular formula does not show the structure of the molecule. Scientists often use structural formulas to illustrate the number and arrangement of atoms in compounds. In Figure 2.1 the molecular formula for methane (CH_4) and ethane (C_2H_6) are illustrated.

Although structural formulas can be very helpful they do not present a complete picture of a molecule. Structural formulas tell us nothing about the distances

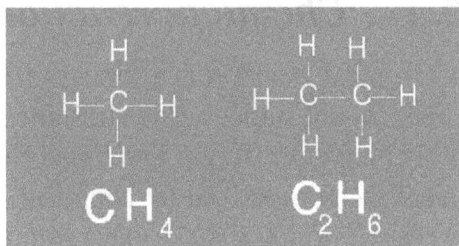

Figure 2.1: Molecular formula methane (left) and ethane (right)

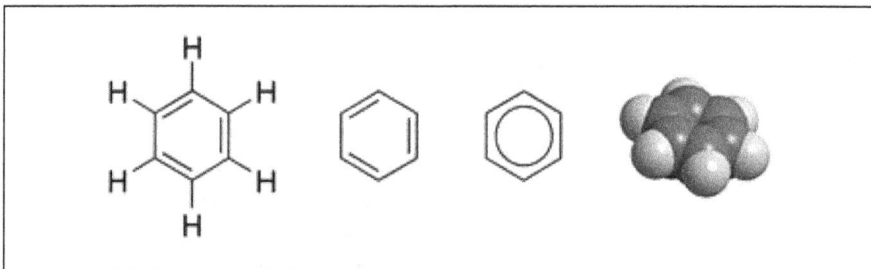

Figure 2.2: Chemical figures of benzene

between bonds, the angles formed by these bonds, or the size and shape of the molecule.

In chemical books the formula of alcohol is written: C-C-OH (ethanol), which means two carbon atoms (ethane) and a hydrogen-oxygen group (alcohol). An organic acid molecule is bonded on two carbon atoms with double bonds (ethylene): C=C-COOH. The acid-group is COOH and can be written as CO_2H. In some cases it is noted as $-\overset{|}{C} = O$ because the bond "=" is a double bond and "-" is a single bond. OH

This chemical acid C=C-COOH is called acrylic-acid and is the raw material for the binder poly-acrylics for acrylic paint.

All aliphatic hydrocarbons are linear carbon chains. The aromatic hydrocarbons are ring structures of six carbon atoms, such as benzene or toluene, with double bonds, also noted with three stripes or a circle in the middle.

In the following Figure 2.2 the six-aromatic ring of benzene has been sketched in three different ways. The fourth structure illustrates the visualisations of benzene with six carbon and six hydrogen atoms.

2.3 *Chemical reactions*

A chemical reaction is a process that always results in a change in chemical substances. The substance or substances initially involved in a chemical reaction are called reactants. Chemical reactions are usually characterized by chemical changes that yield one or more products, which usually have different properties to the reactants. Chemical reactions encompass changes that strictly involve the motion of electrons in the forming and breaking of chemical bonds.

Different chemical reactions are used in combination with chemical synthesis in order to obtain a desired product. The large diversity of chemical reactions and approaches to their study results in the existence of several concurring, often overlapping ways of classifying them. Below are examples of widely used terms for describing common kinds of reactions.

Direct combination or synthesis, in which 2 or more chemical elements or compounds unite to form a more complex product: $N_2 + 3H_2 \rightarrow 2 NH_3$ (ammonia)

Chemical decomposition or analysis, in which a compound is decomposed into smaller compounds or elements: water: $2 H_2O \rightarrow 2 H_2 + O_2$

Displacement or substitution, characterized by an element being displaced out of a compound by a more reactive element: $2 Na + 2 HCl \rightarrow 2 NaCl$ (salt) $+ H_2$

Acid-base reactions, broadly characterized as reactions between an acid and a base, such as HCl (hydrochloric acid) + NaOH (lye) $\rightarrow$ NaCl + H_2O

Oxidation reaction of metals with oxygen from the atmosphere: $Fe + O_2 \rightarrow FeO_2$

The rate of a chemical reaction is a measurement of how the concentration or pressure of the involved substances changes with time. Rates of reaction depend basically on:

- **Reactant concentrations**, which usually make the reaction happen at a faster rate if raised through increased collisions per unit time.
- **Surface area** available for contact between the reactants, in particular solid ones in heterogeneous systems. A larger surface area leads to a higher reaction rate.
- **Pressure**: by increasing the pressure, the volume between molecules is decreased. This increases the frequency of molecule collisions.
- **Activation energy** is defined as the amount of energy required to make a reaction start and continue spontaneously. Higher activa-

tion energy implies that the reactants need more energy, than a reaction with lower activation energy, to start with.

- **Temperature**, which, if raised, speeds up reactions, since a higher temperature increases the energy of the molecules, creating more collisions per unit time,
- The presence or **absence of a catalyst**. Catalysts are substances which change the pathway (mechanism) of a reaction which in turn increases the speed of a reaction by lowering the activation energy needed for the reaction to take place. A catalyst is not destroyed or changed during a reaction, so it can be used again.
- For some reactions, the presence of **electromagnetic radiation**, most notably ultraviolet, is needed to promote the breaking up of bonds to start the reaction – this is particularly true for reactions involving radicals.

Chemical energy is part of all chemical reactions. Energy is needed to break chemical bonds within the starting substances. As new bonds form in the final substances, energy is released. The comparison of the chemical energy of the original substances and the chemical energy of the final substances can be decisive, if the energy has been released or absorbed in the overall reaction. A chemical reaction in which energy is released is called an 'exothermic' reaction. Exo means "go out" or "exit". Thermic means "heat" or "energy". Exothermic reactions can produce energy in several forms. If heat is released in an exothermic reaction, the nearby matter will become warmer.

In organic synthesis, organic reactions are used in the construction of new organic molecules. The production of many man-made chemicals such as plastics, polymers and additives depend on organic reactions. The oldest organic reactions are combustion of organic fuels and saponification of fats to make soap.

3   *Paint ingredients*

Almost all types of paint consist of viscous binders that enable the application of a thin layer (a coating) for adhesion and protection. Pigments are also necessary in order to provide colours in the paint. Many additives and solvents are added to enable good application and stability properties. Additives, such as driers, surface-active additives, biocides etc. are necessary for improving various such as the flow, smoothness, hardness, fouling etc.

Developing and manufacturing paint can be compared with cooking a good meal. The cook invents his own formulas and recipes with his selected ingredients: vegetables, rice, potatoes, meat or fish and various chosen additives: herbs, pepper, salt, garlic etc. There are no two cooks in the world who have the exact same formula or recipe and presentation. Everyone cooks in a different way, using different methods and yet all eaters enjoy the different meals, whether it is an exquisite dinner in a small bistro or a grand buffet in a large renowned restaurant. The atmosphere and surroundings are also very important to making a meal a success. The same comments can be applied to paint. The colour choice and the application conditions influence the paint result: the coating experience. Also the discussion regarding the best paint in the world is the similar to discussing the best cook in the world. Your mother is the best cook in the world and you can eat just as well in both small and large restaurants. The big paint companies can be just as good as smaller paint companies. Smaller paint companies sometimes specialize and make their own special products similar to the comparison we make between a Chinese restaurant and an Argentine restaurant – both different but both just as good! The differences between big and small companies are the logistics, the distribution and the size of the budget for marketing and advertising. The discussion on the quality of paint is all about good ingredients and good formulas.

Adrie Winkelaar: Coatings Basics
© Copyright 2009 by Vincentz Network, Hannover, Germany
ISBN 978-3-86630-851-0

3.1   Binders

Binders are film formers that enable bonding between the pigments and between the surface and the pigments. A coating is a dried layer like a film. The most important binders are

- oil,
- alkyds,
- acrylics,
- epoxies and
- polyurethanes.

All these substances have different properties in adhesion, hardness, flexibility, gloss retention, water resistance and UV-resistance. The question is, how it is possible and what causes these differences?

There are natural binders, such as linseed oil and resins, and also modified natural substances such as nitrocellulose, alkyd and synthetic substances such as acrylics, epoxies and polyurethanes.

In each coating of applied paint a transition from liquid to a solid state takes place. A distinction is made between **physical drying** and **chemical curing**. This can occur simultaneously or in sequence.

During the physical drying of an applied paint, the transition from liquid to a solid state is the result of evaporation of solvents, including water. An example is the old-fashioned cellulose lacquer used in former days. Another form of physical drying is coalescing drying of waterbased dispersion paints. The binder particles flow together after the evaporation of the water. During the chemical curing of an applied paint, the transition from liquid to a solid state is accompanied by an increase in molecular mass. Examples are two-component paints, which must be mixed together before application. The oxidative drying paints such as the well-known oil or alkyd resin have oxidative drying by oxygen from the air. See Chapter 4 for drying mechanisms of binders.

3.1.1 Natural binders

In the present paint industry natural substances such as resins, plant oil and bitumen are still used. They must be diluted with solvents such as white spirit, xylene, alcohol or acetone.

Resins

The most important natural resins are colophony, copal, dammar and shellac. Colophony is extracted from pine trees and is diluted in white spirit. This resin has a high acid value and has to be combined with oxides of zinc, magnesium or calcium to neutralise it. Shellac is the product of the sap of certain trees and is produced by the insect called Laccifer lacca commonly known as the 'lac insect'. This amber coloured resin is soluble in alcohol and can be made water-soluble in combination with amines or ammonia. Resins provide a hard film and are combined with other binders.

Plant oils

Important oils in paint chemistry are linseed oil, soya bean oil, safflower oil and castor oil. Linseed and soya bean oil are drying oils, safflower oil is semi drying and castor oil is non-drying. The fatty acids in the oil are, together with other reactive groups, responsible for drying properties. Through an oxidative reaction the oil changes into a soft and sticky substance. The drying oils play an important part in modified binders such as alkyds. In Table 3.1 the compositions of some plant oils are described.

Table 3.1: Composition of some plant oils

	Linseed oil	Safflower oil	Wood oil	Soya oil
Palmitin acid	6.0	6.0	4.0	10.0
Stearin acid	4.0	2.0	1.0	3.0
Oil acid	19.0	19.0	4.0	29.0
Lineic acid	24.0	70.0	-	51.0
Linoleic acid	47.0	3.0	1.0	7.0
Eleostearin acid	-	-	90.0	-
Total weight %	100.0	100.0	100.0	100.0

Bitumen

Bitumen, asphalt and pitch are dark, hard substances that contain excellent water resistant properties and are used for corrosion protection, roof coatings and vehicle under seals. They are combined with other binders into one and two component products. The use of coal tar is forbidden because it contains toxic polycyclic aromatic hydrocarbons (PAC).

3.1.2   Modified natural binders

Modified binders play a big part in natural substances. Cellulose, rubber and fatty acids (plant oils) can be modified to become good binders.

Fatty acids: alkyds

A combination of "alcohol" and "acid" is called **alkyd**. In chemical terms this reactive product is called **polyester**. The structure of an alkyd consists of oil, alcohol (glycerol) and phtalic acid. More phtalic acid in the alkyd presents a harder and faster drying binder. Increased plant oil in the alkyd results in a weaker and slower drying binder. A high fat-content binder is for exterior application and a low fat-content binder is for interior application.

An alkyd is formed in three steps by

- phtalicacid: $C_6H_4(COOH)_2$, a benzene ring with two acid-groups,
- glycerol: $C_3H_5(OH)_3$, an alcohol with three OH-groups
- carboxylic acid: R-COOH, which R is a long chain of linseed oil, that is a mix of linolenic acid, linoleic acid and oleic acid.

There are fatty alkyds containing more than 55 % of plant oil and lean alkyds containing less than 45 % of plant oil. If the alkyd is fatty, the phtalic acid is much less. See Table 3.2.

Figure 3.1: Alkyd molecule structure

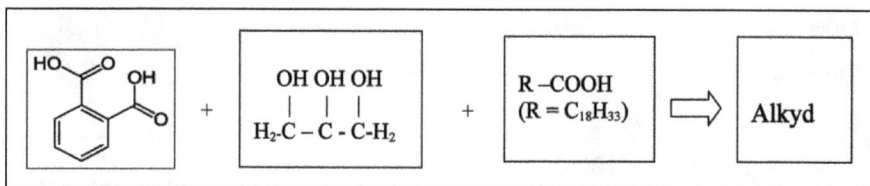

Table 3.2: Types of alkyds

Types of alkyds	Oil content (%)	PA content (%)
Oil free	0	77
Lean	<45	>40
Medium fatty	45 to 55	30 to 40
Fatty	>55	15 to 30

If the oil content is high, then the alkyd viscosity is low and the brush ability is good. If the viscosity of a lean alkyd is too high then the brush ability is bad. The thickness of a dried layer from a fatty alkyd is more than the thickness of dried layer from a lean alkyd. The flexibility of a fatty alkyd is higher than that of a lean alkyd. The wet paint film-flow is much better from a fatty alkyd. The yellowing of a dried fatty alkyd paint film is increased compared to a lean alkyd paint film. The durability of a medium fatty alkyd is the best.

Alkyds conforming to the Directive 2004/42/EC (see Chapter 10) have to be composed of more less-solvents. The glycerol changes through an organic alcohol with four or five reactive groups that provide more branching alkyds and better drying time. The low viscosity is attained with shorter oil chains.

3.1.3   *Synthetic binders*

The most famous synthetic binder is acrylic which is obtained through a polymerisation process of raw materials from crude oil. This means that a long chain of carbon molecules is built up from the reaction of each molecule together with another molecule. In addition, epoxy and polyurethane binders are now being manufactured.

Acrylic

The acrylic "monomer", one molecule of pure **acrylate**, comes from the petrochemical industry and can be combined with other monomers. The chemical structure is based on two carbon atoms and carbon acid. A double connection between two carbon atoms is very reactive and can react with other monomers to a polyacrylic

$$C = C - COOH$$

The size of these new molecules is expressed in molecular weight and can range from around 1000 to 3000 g/mol. In practise, it is possible to create many different types of acrylics, such as polyacrylate, methacrylates and styrene acrylates. Each synthetic polymer has a glass transition temperature, which means that the higher the temperature, the harder and less flexible the binder film will be. The glass transition temperature rises as a result of chemical cross-linking. A polymethacrylate has a higher glass transition temperature than a pure polyacrylate.

Acrylic resins are characterised by good chemical and photochemical resistance. This means there is no yellowing and a low UV-degradation in the coating. Acrylics polymers can be solved in solvents and can be dispersed in water (see Chapter 4).

Epoxy

In the organic chemical science there is a special three-member ring resulting from two carbon atoms and one oxygen atom. This is known as the "epoxy group".

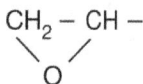

$$CH_2 - CH - $$
$$\diagdown \, O \, \diagup$$

Each epoxy resin has at least two epoxy groups per molecule and is linked to a "bisphenol group". Bisphenol is composed by two aromatic rings of phenol (C_6H_6-OH). Epoxy groups have a tendency to react with amines, so that a two-component product has to be made. Before application both components must be mixed together, after which, the reaction begins immediately. An epoxy reactive molecular group is expressed in equivalent weight (g/mol) and must be linked to the same equivalent weight of an amine for optimal reaction. When too less or too much equivalent weight is present in each component, the remaining component will negatively influence the properties.

Epoxy resins came on the market after the Second World War and have particularly strong properties such as excellent strong adhesion, excellent water resistance and water stability. For this reason epoxies are used in primers and for metal protection.

Polyurethane

A polyurethane is also a well know polymer based on a chemical reaction from "isocyanate" together with an alcohol-group. An isocyanate group is built up from one nitrogen atom, one carbon atom and one oxygen atom.

$$- N = C = O$$

This means that polyurethane is a two-component product. Binders contain a lot of the OH-groups. The harder an isocyanate this means it must be mixed before the application. Because an alcohol group resembles a water molecule, polyurethane is also produced as moisture curing quality product. This resin belongs to the one of the component systems. Skin contact with polyisocyanates and especially inhalation of spray mists must be avoided in all circumstances.

The polyurethanes are both thermally and chemically very stable and have a high gloss performance. The first applications, produced by Bayer in Germany, were also used after the Second World War. These coatings are sometimes known as DD coatings, which represent the first two letters of the trade names of Bayer's polymers.

Similar to the epoxy binders, the equivalent weight (g/mol) of the isocyanate and the OH-groups must be calculated for the correct ratio.

Silicone resins

Silicone resins belong to the class of silicones (Si = 14, number in the periodic table (see Appendix 1)) with the scientific name "polysiloxanes" whose characteristic structural feature is the Si - O - Si chain. Many silicones are incompatible to other polymers. Reactive silicones can be linked with alkyds (silicon modified alkyds), polyesters (silicone polyesters) and acrylic resins. The use of silicone resins improves heat resistance, water resistance, surface smoothness and UV retention (gloss maintenance).

Silicate paints, based on potassium water glass, have film forming caused by absorption of atmospheric carbon dioxide with the precipitation of polysilicis acids, which eventually convert to silicon dioxide (sand). The characterised property is the extremely hard film forming that is resistant to weathering, light and chemicals. The coating is permeable to gas and water vapour.

Mixtures

Not only the chemical industry, but also the paint industry itself is increasingly experimenting mixing different binders together in order to reach a special quality. There are two mixes with possibilities: namely "hybrid" and "modification" mixtures. A **hybrid** is a mixture, for example, of an alkyd solution or emulsion (see Chapter 3) and an acrylic dispersion. The two binders have different consistencies, namely, white spirit solution and water-based dispersion. The alkyd binder is flexible and oxygen drying (thermo hardening) and acrylic is a UV resistant clear binder with physical drying properties (thermoplastic). A thermoplastic binder becomes sticky when heated and sandpapering of the dried layer is not good. A thermo-hardening binder has chemical drying and hard coating properties, also when heated.

Table 3.3: Some properties of modified alkyds

	Standard alkyd	Acrylic mod. alkyd	PU mod. alkyd	Epoxy mod. alkyd	Silicon mod. alkyd
Drying time	+	++	++	+	+
Adhesion	+	+	+	++	+
Hardness	+	++	++	+	+
Durability	+	++	+	+	++

A modification is a chemical bonding of two different binders. There are alkyds with acrylic connections (acrylic modified alkyd), epoxy modified alkyds, urethane modified alkyds and silicon modified alkyds. All these alkyds are oxygen drying binders and soluble in white spirit. The properties of these alkyds have somewhat the origin of modified binders. See Table 3.3.

3.2 Solvents

A solvent is a liquid, consisting of one or more components, and is able to dissolve binders without the need of a chemical reaction. Most solvents are volatile under normal application conditions and make paint processing suitable for application. How can solvents dissolve binders and how many solvents are used in the paint industry? The most important solvent groups are aliphatic hydrocarbons, aromatic hydrocarbons, ketones and esters.

3.2.1 Aliphatic hydrocarbons

Aliphatic solvents are different types of white spirit (turpentine). The chemical name of aliphatic hydrocarbons means that all the contained carbon atoms are in long chains. These aliphatic solvents dissolve mineral oil, plant oil, waxes and paraffin. The most important physical properties are:

- boiling point in °C,
- flashpoint in °C and
- solvability.

The solvability of a solvent is based on the destruction of the molecule bonds in the substance to be dissolved. There are three different

Table 3.4: Some properties of modified alkyds

Aliphatic hydrocarbon	Solvability	Boiling point [°C]	Evaporation rate
Hexane	7.3	69	4.1
White spirit	7 to 8	150 to 190	85 to 90
Methanol	14.5	65	6.1

forces between molecules, namely the dispersion or London-forces, the dipole forces and the hydrogen forces. All solvents have a different mixture and can be calculated by the parameters of *Hansen* and *Hildebrand* for the different solvability.

3.2.2 Aromatic hydrocarbons

The most well known aromatic solvents are toluene and xylene. These are generally more expensive than aliphatic solvents and have a higher dissolving power. The carbon atoms are in hexagonal rings and have a strong penetrating smell (aroma). They can dissolve alkyd resins, polyesters, acrylates and many other substances.

3.2.3 Ketones and esters

Ketones such as acetone and "methyl ethyl ketone" (MEK) can also dissolve numerous binders such as nitrocellulose coatings and oven-drying coatings.

Table 3.5: Properties of aromatic hydrocarbons

Aromatic hydrocarbon	Solvability	Boiling point [°C]	Evaporation Rate
Toluene	8.9	110	2.4
Xylene	8.8	140	0.7

Table 3.6: Properties of ketones, esters and water

Ketones and esters	Solvability	Boiling point [°C]	Evaporation rate
Acetone	9.8	56	11.6
MEK	9.3	80	5.7
Ethyl acetate	8.8	77	1.0
Ethyl-glycolacetate	9.7	155	0.2
Water	23.4	100	0.1

Esters have the same characteristics as ketones, but generally have a more pleasant odour (fruity). Ethyl acetate is an important solvent for quick drying coatings such as nitrocellulose and polyurethane products. The standard of the evaporation rate is ethylacetate to the value of 1.0.

Water is becoming more and more the most important solvent in the paint industry. Water has a very low rate of evaporation and can be mixed with alcohol, ketones and some esters. The surface tension of all binders and solvents varies. Water has a high surface tension (about 70) and white spirit a very low surface tension. Surface tension is determined by cohesive forces and causes the well-known water-drops on fatty surfaces. When ammonia is added to water the drops disappear because the ammonia decreases the cohesive forces. This is the why fatty surfaces need to be cleaned with ammonia or soap products before applying a water-based paint.

3.3   Pigments and extenders (fillers)

A pigment is a substance that consists of particles and is practically insoluble. It is used as a colorant. There are also corrosion inhibiting pigments and special effects pigments, such as those containing magnetic properties and others having mother-of-pearl effects. Dyes, in contrast, are colorants that are soluble in an application medium, such as water, alcohol and other solvents. Extenders (fillers) are substances consisting of insoluble particles that increase the pigment volume in paint enabling better filling properties that influence optical properties, such as gloss. The cost of extenders is also very inexpensive and is often used to reduce the total raw material price of the paint. From a colouristic point of view, pigments are divided into organic and inorganic types. It is possible to make all the colours of the rainbow with the use of organic and inorganic pigments.

Pigments and extenders are never used separately; they are always combined in a paint formula. The interaction between the particles and the medium is important for the optimal effect. The size and the surface of the particles determine the properties of the entire system.

The size of the surface area of a given amount of particles depends on its density and particle size distribution. The lower the density is and the smaller the particles are, the larger the surface area becomes. One cm^3 of particles with a diameter of $1\,\mu m$ (one micron) has a surface area of $6\ m^2$. Particles with a diameter of $0.1\,\mu m$ have a surface area of $60\ m^2$. Very fine particles have a better hiding power and give a more intensive colouring.

3.3.1   Inorganic pigments

Inorganic pigments belong to the group of oxides. The most important oxides are **titanium oxide** (white) and **iron oxides** such as yellow, red, brown and black. In the past these pigments were found in the ground, but nowadays they are made synthetically. Natural oxide pigments such as ochre, amber and burnt sienna are used in small amounts for manufacturing art painting products. They have different compositions with iron oxides, combined with other metals like manganese (number 25 in the periodic table of elements, see Appendix).

The application of heavy-duty pigments, such as chromate, molybdate and cadmium pigments has ceased because of their toxic poisons they are hazardous to health and very bad for the environment.

The speed towards water and light is independent of the particle size. On the contrary, hiding is very much dependent on the particle size. Very fine inorganic pigments with particles $<0.05\,\mu m$ are transparent. Coarse pigments with particles between 0.1 and $1\,\mu m$ have excellent hiding powers. In contrast, particles with more than $10\,\mu m$ display relatively poor hiding powers.

Table 3.7: Differences between rutile and anatase

	Rutile	Anatase
Refractive index	2.75	2.55
Density	4.1 g/cm³	3.9 g/cm³
Light absorption	<415 nm	<385 nm
Hardness (Mohs)	6.5 to 7	5.5

The most important white pigment is titanium dioxide. There are two different types of titanium dioxide – anatase and rutile. These white pigment types have different crystal structural properties. Rutile is of greater importance for coatings, because it is harder and more stable, see table 12.

Iron oxide pigments also contain different crystal structures. Moreover, iron oxides have different chemical formulas. Iron (Fe) has different oxygen (O) connections that produce different formulas. Fe_2O_3 is the well-known red brown colour and Fe_3O_4 is black iron oxide. FeO is a yellow formula with a special needle-like crystal structure.

Pigment particles can be recognised as individual entities through suitable physical processes e.g. using microscopes. Since particles or aggregates cannot be further divided into individual particles under the shear force that occurs during dispersion, individual particles are occasionally primary particles. The smallest entity that occurs in pigments is a crystal.

3.3.2 Organic pigments

Since the use of heavy-duty poison inorganic pigments has been restricted, the introduction of organic pigments is increasing. Organic pigments are complex hydrocarbon structures within the chromofore chemical groups. The most important organic pigment is phtalocyanine (blue and green) containing copper atoms (Cu) within the crystal structure. Organic pigments can be divided, more or less, into three groups: namely azo-pigments, polycyclic pigments and metal complex pigments such as phtalocyanine.

Azo pigments

The mono-azo and di-azo pigments consist of a general characterized azo-formula "– N = N –" in their structures. Mono-azo pigments have one group per molecule and di-azo two groups per molecule. Azo pigments cover the red to yellow colour range. The light density performance of the azo-pigments in contrast to the inorganic pigments is moderate.

Polycyclic pigments

Within the group of polycyclic organic pigments, there is a diverse range of substances present. Phtalocyanine also has a polycyclic structure; however, copper atoms present the classification as metal complex pigments. Indoline pigments are yellow, orange and red. Dioxazine is a brilliant violet coloured pigment

Metal complex pigments

Phtalocyanine pigments are green and blue. The blue tinted pigments become green after adding chlorine (Cl) atoms. The more chlorine-atoms, the more the yellow tint appears and blue then becomes green. Together with the light density performance, which is relatively good, and the low costs, these organic pigments are broadly used for colour matching application.

3.3.3   Extenders

In former times several cheap substances were added to paints in order to obtain more paint materials. This activity resulted in the name 'extender' or 'filler' in modern paint formulas. Numerous extenders were added, especially to wall paints, for coating large interior and exterior surfaces of houses and other buildings. An extender is a substance consisting of particles that are practically insoluble in the medium (binders and solvents) and are used to increase volume, to improve technical properties and/or to influence optical properties.

Extenders can be categorised in particles size. Coarse grinds with a particle size above 250 µm for plasters and concrete fillers.

Medium size particles, between 50 and 250 µm, are used for creating good particle size distribution. If the particle size distribution is unbalanced, cracking in the dried film/layer can occur.

Fine extenders have a particle size between 10 and 50 µm and ultra fine extenders have a particle size of less than 10 µm. The last mentioned can be used as flattening agents in order to reduce the gloss in a coating.

Extenders can be classified into five groups:

- carbonates,
- silicon dioxides,
- silicic acids,
- silicates and
- sulphates.

Carbonates

Chalk, calcite and dolomite are the three most important carbonates. The chemical formula is $CaCO_3$ and is a natural substance. Chalk is only used in products for interior use. Calcite has a crystal structure with good all-round properties. The most important applications are in primers, wall paints, fillers and silky gloss paints. Dolomite is a mixture of calcium and magnesium carbonates that contains somewhat harder properties and is less sensitive to acids. The name derives from where this substance is found.

Silicon dioxide and silicic acids

These substances are based on SiO_2 and have different crystal structures. Some substances are natural, such as kieselgur and diatomaceous earth, but other silica's are manufactured as silica flour and pyrogenic silica's. Pyrogenic silica's consist of very fine particles and are used as thixotrophic or anti-settle agents. The particles have diameters of between 5 and 50 nm. The precipitated silicic acids are used as flattening agents to reduce the gloss in paints. A fine crystal silicon dioxide fraction can give lung embolism or lung cancer after inhaling

Silicates

The most important silicates are talc, kaolin (china clay) and mica based substances. These substances also have different crystal structures. Talc is a magnesium silicate ($Mg_3Si_2O_{10}$) with a particular layer structure with a very high oil absorption value. This is the quantity of oil necessary to wet a standard quantity dried particles. The chemical inertia is the basis for use in anti-corrosive primers, however the hardness is minimal (Mohs' value is 1). For this reason it is used in fillers, because it is very good for sandpapering.

Kaolin is an aluminium silicate with good weather resistance properties. Mica is also an aluminium silicate with a special crystal structure and excellent chemical and heat resistance. The crack-bridge action of mica when added to coatings is also well known. Both are used in primers, wall paints, fillers and anti-corrosive systems.

Sulphates

The two sulphates usually found in paint are 'blanc fixe' and barites. Barium sulphate ($BaSO_4$) has a very high density so that the weight of a litre product is also very high. 'Blanc fixe' is a synthetic type and barites have a natural quality. Both have a narrow particle size distribution of approximately 1 µm.

3.3.4   Special pigments

Anti-corrosive pigments, **special effects pigments** such as a 'lustre' effect and **dyes** are based on metallic pigments, which are generally platelet-shaped particles of non-ferrous metals. Zinc and lead are the most usual metallic pigments containing anti-corrosive properties and aluminium (flakes) is mostly used for creating lustre effects.

An anti-corrosive pigment is understood to be a pigment that, when used in a primer on metals inhibits or prevents the corrosion of the metal surface. This is generally due to a chemical or physics-chemical reaction. Corrosion is an electrochemical process caused by oxygen and water on metal. Corrosion stimulators such as chloride or sulphate molecules can accelerate this process. The anti-corrosive pigments intervene on this process. An alternative method to protect metal surfaces is by applying a thick and dense layer. Anti-corrosive pigments also extend the diffusion distance of water and aggressive substances such as oxygen or salt molecules.

Zinc phosphate is the most commonly applied anti-corrosive pigment. In former times red lead, lead chromates and zinc chromates were used.

Aluminium is the grinding of aluminium platelets (flakes) into different qualities and then applied as a physical barrier against water, reflection of heat and UV radiation. The reflection of light on a surface causes the metallic effect. Bronzes, coppers, gold, silver and nickel alloys are also used.

Natural dyes, such as indigo or purple have been well known for many years. Synthetic dyes are divided into metal complex dyes, anionic dyes and azo dyes. The paint industry uses these in very small amounts for the creation of special effects.

3.4 Additives

Paint contains numerous additives. An additive is a substance added to paint in small quantities in order to impart specific properties both in wet paint and the dried layer (coating). They are also called auxiliaries. Additives are generally divided into various groups such as: defoamers, wetting and dispersing agents, surface active additives, flattening agents, rheological additives to influence viscosity, corrosion inhibitors, light stabilisers, driers, accelerators and biocides.

3.4.1 Auxiliaries in wet paints

To make a stable paint, an **anti-settle agent** is necessary in order to prevent separation between pigments and binders during storage. We have learned that pigment particles must be dispersed in the binder. Wetting and dispersing agents are necessary to make a stable liquid. Most of these additives are based on surface tension agents such as soap.

During application the paint must flow like water on a flat surface. Flattening and **defoaming agents** are added for the realisation of this effect. There is a different between water-based paints and solvent-based paints. The surface tension of both types is very different. In aqueous systems mineral oil defoaming agents and silicone defoaming agents are used and in solvent containing products polysiloxanes are used.

The **rheological additives** are important for good performance during application. The paint must be applied in the required layer thickness and must flow well without dripping. **Thickeners** are designed to prevent pigments from setting out during storage and transportation. However, too much thickener influences the application. For water-based paints the well-known cellulose thickeners, also known as wallpaper glue, are used. More often acrylic and polyurethane thickeners are increasingly being used. Pyrogenic silicic acids and hydrated castor oil are used for solvent content products. **Thixotropic agents** (see Chapter 4.2) provide special effects with rheological properties. To prevent dripping the viscosity must be high, but for a good flow the viscosity must be low. Thixotropic agents present this effect.

For aqueous materials a **biocide** is necessary to stabilise the medium against bacteria. Bacteria in storage containers leads to fouling which in turn will change the rheological properties. During this process a putrid odour and gas evolution are formed. To avoid this in-can situation, preservatives are necessary.

3.4.2   Auxiliaries in dried coating

For dried coatings **light stabilisers** can be added. In opaque colours they are not efficient, but in clear varnishes they provide long-term lasting performance. UV absorbers are active because of the fact that harmful UV radiation in a wavelength range from 290 to 350 nm is absorbed and converted into harmless heat energy.

Catalysts and driers are necessary to accelerate the cross linking reaction in a chemical active binder. The oxidative drying oil and alkyd resins must have "siccatives" in order to start up the chemical reaction. Cobalt (Co), manganese (Mn), zirconium (Zr), calcium (Ca) and barium (Ba) are substances that are used in very small quantities. These materials have an inherent catalytic action. Salts of calcium and barium are only slightly active in combination with the other metal salts. Owing to ecological and toxicological reasons the use of lead is now avoided wherever possible. Other catalysts, which accelerate the cross linking of the binders, are used in stoving coatings, quick-drying coatings and also acid-curing coatings.

Biocides are used as a film-preservation additive, to protect the applied coating against bacteria, mould, algae or moss forming. Such protection is principally required if building materials and coatings come into contact with moisture for extended periods. Each anti-fouling product has to be registered in each EU-country before market introduction.

For smooth surfaces special additives such as silicon and Teflon are added. For anti-slip properties coarser and bigger particles are added.

4   *Consistency and stability*

Since almost all paints are liquid products, binders can be applied in different viscous consistencies, such as solutions, dispersions and emulsions. Solutions are obtained through solvents. See Chapter 2. Dispersion is a fine distribution of solid particles in a liquid. Pigments, extenders and solid binders can form dispersion. An emulsion is fine distribution of liquid particles within another non-mixture liquid. An example is an emulsion of oil or fat fluid particles in water.

The different consistency technologies are also used in the food and cosmetic industries. Milk is an emulsion of liquid fat in water. Butter and other dairy products are also emulsions. Many creams and body milks in the cosmetic industry are emulsions. Questions in this Chapter 4 are how to make emulsions and dispersions?

Two different liquids such as oil and water can be mixed together, but how is a stable emulsion made? Some knowledge regarding emulsifiers and dispersing agents is necessary to be able to understand the answers to these questions.

A lot of paint research is necessary in order to create a stable consistency of all the particles contained within paint, so that it can be stored in factories and shops for long periods of time. The manufacturing of solid paint is also possible. Powder coatings are applied electro statically and the results, after being dried at high temperatures (stoving) are very strong layers of coatings. Increasingly more high solid and solid paints are being developed to answer to the high-demand pressure for solvents in our present society. Solvents are volatile and dangerous to the health and environment. In Chapter 10, the various legislations for Europe and member states that deal with and pressurise the laws and regulations regarding solvent application in paints, are summarized.

Adrie Winkelaar: Coatings Basics
© Copyright 2009 by Vincentz Network, Hannover, Germany
ISBN 978-3-86630-851-0

4.1 Solutions, emulsions and dispersions

The traditional solutions, based on solvents, are rapidly disappearing. The paint industry is developing and introducing increasingly more and new different consistencies.

4.1.1 Binder solutions

The purpose of solvents is to dilute a binder and adjust the viscosity for application. Like solvents, binders also have solvability parameters, see Chapter 2.2.1. The *Hansen* and *Hildebrand* parameters of binders can be calculated and can be expressed in solvability values. See Table 4.1. If the same value of a solvent is higher than the binder value then the solvability will be very poor. A selected solvent having lower values presents good solvability.

Solvent mixtures are also applied because in some cases solvent mixtures are better than pure solvents. This theory is important because there are fewer solvents needed when the solvability is in balance using correct combination.

A solution is a binder concentration in solvents, such as oil or alkyd in white spirit. All oil and alkyd paints are dissolved in white spirit. All epoxy and polyurethane paints are dissolved in aromatic hydrocarbons and esters, such as ethyl acetate and are grouped together under the general name of "thinner". In general, all world wide paints can be recognized by these three binder solvents: water is dispersion paint, white spirit is always oil or alkyd based paint and thinner is an epoxy or polyurethane based paint.

Table 4.1: Solvability of some binders

Binder type	Solvability
Cellulose nitrate	11.3
Fatty oil alkyd	10.4
Epoxy	12.8
Petrol resin	8.8

4.1.2 Dispersions

Dispersion is the fine distribution of solid particles in liquids that cannot be dissolved in the liquid, such as acrylic particles in water or pigment particles in oil.

During the manufacturing of dispersions there are many various degrees of fineness possible. The first acrylic dispersions in water were coarse with a fineness of 10 µm. Later 1 µm particle dispersions could be manufactured resulting in very fine particles of 0.1 µm. In Chapter 6 some of the grinding methods of pigment dispersions are described.

In modern binder dispersion techniques increasingly more and different binders are possible. Polyurethane dispersions in water are also being manufactured. With these dispersions very hard coatings can be created, such as varnishes for parquet floor covering.

A modern technique is the creation of an ultra fine dispersion with minute particles (nano technology). A nanometre is one thousandth of a micrometre and one micrometer is one thousandth of a millimetre. This ultra fine technology presents new possibilities. The specific surface of these particles is enormous. If a particle is 1 cm cube the surface will be 6 cm². In one cm³ volume 1012 particles of 1 µm can be filled together totalling a specific surface of 6 m². One cm³ nano particles has a specific surface of 6000 m². This is more than the measurements of one football field.

The dispersion stability is very important. Dispersing agents provide this stability through the electrostatic and steric effects between the molecules. There are four groups of dispersing agents: non-ionic agents, anionic agents, cationic agents and amphoteric agents.

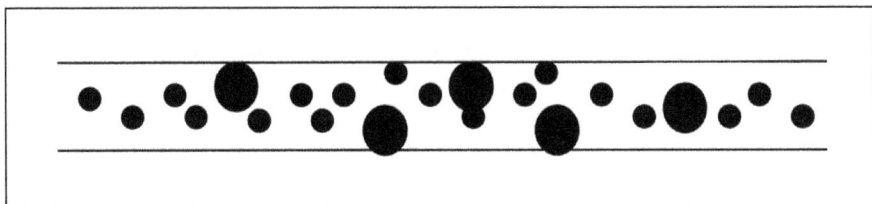

Figure 4.1: Fine distribution of solid particles in a thin layer on a substrate

Non-ionic agents have hydrophilic and hydrophobic groups, such as soap. The hydrophilic side of the molecule enables binding with water, and the hydrophobic side enables electrostatic binding with solid or liquid particles. Examples are ethoxylates and lecithins.

Anionic agents are sodium salts of organic compositions. The electrostatic binding is formed through the positive charge of sodium and the negative charge of liquid or solid particles. Cationic agents are based on a negative charge, such as chlorinated or fluorinated organic salts. The liquid or solid particles have a positive charge.

The amphoteric agents have both anionic and cationic properties. If the pH (acid grade) is high, then the amphoteric agents are anionic and if the pH is low there are cationic properties.

The dispersing agents, also called **wetting agents** or **surfactants**, influence the surface tension of a solvent or a solution. Hydrocarbon surfactants decrease the surface tension of water from 73 down to 25 to 40 units in mN/m. Silicone surfactants decrease water to 20 within 30 units and fluorinated surfactants decrease water to 15 within 25 units.

Modern binder dispersions also have complex structures such as the "core-shell" structure. The particles consist of two different polymer substances, for example, an elastic inner structure and a hard outer shell structure.

4.1.3   Emulsions

An emulsion is the fine distribution of one liquid into another liquid that cannot dissolve together. The two liquids are intolerant, like oil

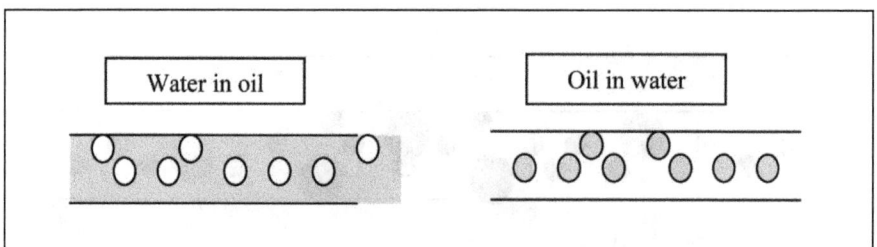

Figure 4.2: Thin layer of water in oil and oil in water

and water. In this case there are two possibilities: oil can be emulsified in water and water can be emulsified in oil. A water-in-oil emulsion can be diluted with white spirit and oil-in-water can be diluted with water.

In this way an alkyd can be emulsified in water and need less solvents. Two important problems may arise: the stability of the emulsion and the drying stability of the alkyd. During the emulsification process a good emulsifier (surfactant) is necessary, after which, a good stabilizer for the emulsion. The two phases may split up at any time and it is extremely difficult to remix.

An alkyd emulsion can be created by heating pure alkyd to 85 °C and adding a non-ionic agent. During the mixing, warm water is added to the alkyd. The drying stability can be improved with the use of special driers suitable for water-based alkyd resins. It is becoming increasing and possible for different binders to be emulsified in water such as epoxy, chlorinated rubber and cellulose nitrate.

4.2   Viscosity and thickeners

The viscosity of substances normally rises with increasing molecular weight. Monomers have very low viscosity and polymers have a high viscosity. The viscosity is more or less the liquid thickness. The paint viscosity depends on the binder viscosity, and also on the paint composition. Most pigments and extenders increase viscosity and most solvents decrease thickness.

Viscosity

A low viscous liquid streams rapidly through a narrow pipe, but a high viscous liquid would stream through very slowly. Liquids resist against the stream and the measurement of the resistance is the viscosity. The official measurement unit is the Pascal second (Pa·s). An old measurement is called the 'poise' (P). 1 Pa·s = 10 P.

The viscosity is dependent on numerous factors. One important factor is the temperature. The higher the temperature the lower the viscosity. A cold paint at 5 °C would be very difficult to apply, and a warm paint at 40 °C is low viscous and easier to apply. Another factor is the type of binder and type of solvent content. The solvability of both is very important – see Chapters 2.2.1 and 3.1.1.

There are two types of viscosity, namely linear and structure. Linear viscosity is lower with proportionate forces. When stirring a liquid with more forces, the viscosity becomes lower and lower. If the viscosity has a structured consistency, for example, thixotropic, there is no linear connection between the thickness and the forces enabled through mixing. All the properties of application, such as viscosity and the paint flow of the paint film, are "rheological". Rheological additives, such as thickeners, can influence rheological properties.

Thickeners

Thickeners are designed to prevent the setting-out of pigments and extenders in paints during storage and transport. They may also influence many of the application properties. The most important area of application is the water based dispersion paint. The thickening action is based on the thickening of the liquid-phase caused by swelling. The most well known thickening agents are cellulose derivates (e.g. methyl cellulose, hydroxyethyl cellulose), acrylate copolymers and PU thickeners. A three-dimensional network is formed, which can be easily broken down again. Influences on thickeners are the pH (acid value) and the biological degradation by bacteria.

In order to obtain good flow with no dripping, a thixotropic agent can be added. Well known thixotropic agents are phyllosillicates (e.g. bentonite, hectorite), pyrogenic silicic, polyurea derivatives and hydrated castor oil. All these substances build up a colloidal gel structure in water and in solvent content paint.

4.3   High solid paints and solid paints (powder coatings)

In addition to the two paint types such as water-based and solvent containing products, increasingly higher solid and powder coatings are being introduced. High solid paints have a high solid content, so that fewer solvents are necessary. The pressure of demand for solvents is so high that more and more solid paints are being developed. The binder molecules in high solid systems have lower molecular weights than those in other systems. Solvent free systems are possible in two-component products such as epoxy and polyurethane formulas.

The most important solid paint is powder coating. This consistency is different from liquid paint products. At present, with only a few exceptions, all polymers used for paint formulas can be transferred into dry powder that can be sprayed electro statically. Polyester, acrylic, polyurethane and epoxy are available in powder quality.

During the last century in the 1950's, powder coatings were developed to avoid the risk of fire and explosions. At present powder coatings have been introduced for environmental reasons – they are solvent-free and can be recycled. The overspray of powder applications is returned for the process of re-application. For this reason the transfer efficiency is very high and there are no problems with waste materials.

4.4 Stability and other wet properties

The stability of paint is very important because of the many particles together in one composition. Binder particles (dispersions or emulsions) and different pigment and extender particles are in balance after manufacturing and need to remain in balance after months and even years in storage. Therefore good storage circumstances are very important especially for water-based products. If an emulsion or dispersion has been frozen, the balance will be disturbed and the paint will become unusable.

Another stability problem is sediment in the paint can. The pigment and extender particles have a weak composition with the binder and the wetting agents. Stirring before application maybe the solution, however, the paint quality is no longer optimal. If the sediment is too old, the paint is difficult to stir by hand.

Other wet paint properties are: viscosity, thinning ability, paint fineness and solid content. The higher the solid content the better the filling properties are of the dried coating. Layer thickness is very important for anti corrosive paints (steel protection) and good weather resistance coatings. The paint fineness is important for hiding power and colour intensity (see Chapter 3.3).

5   *Coating properties*

Paint has many, many properties of which there are four main ones. The most characteristic property is the **drying time**. During the drying time physical and chemical processes take place. With some binders there is only a physical drying time, while others have a combination of physical and oxidation. Most chemical dried paints have two-component or radiation drying. In addition to drying, each paint or coating has **resistance properties**, e.g. water resistance, scratch resistance, etc. Last but not least, all paint has **durability properties** – after all, paint is **colour** and colour is paint.

People expect the results of a newly applied coating to have a beautiful appearance (colour and gloss) and provide substrate protection against e.g. scratches (short term) and against weather influences (long term). Paint properties link up with the substrate properties such as wood, metal, stone, concrete or plastic. Paint applied to wood should not crack, paint applied to metal should not corrode and paint applied to walls should not cause pollution. The various binders provide several possibilities although each binder has its own typical properties. The new nano technology can retain some substances in particles that are released when needed, for example when damage occurs (self repairing).

Colours are very important to our lives. Our visual senses present visions in our minds that are multi coloured. All colours present different impressions and each colour can be explained by psychologists as presenting experiences such as: a cool, warm, violent, calming, youthful, old, lively, dead, spiritual, sexy, etc. A world without colours is impossible to perceive and those people who are colour blind are presented with great difficulties in life. Colours are used for road signs, for emergency exit signs in buildings, in hospitals, in aeroplanes and at airports – to mention but a few. Colours are the

Adrie Winkelaar: Coatings Basics
© Copyright 2009 by Vincentz Network, Hannover, Germany
ISBN 978-3-86630-851-0

basic substances of art, of our identity, such as flags, clothes and numerous other things. Understanding how colours are composed from pigments and dyes in colour tones (hue), saturation and brightness, will supply a broad knowledge of all applications. In this chapter, the most important coating properties are explained.

5.1 Drying

In Chapter 2.1 three types of drying were described: physical drying, coalesce drying and chemical drying such as two components, oxidative drying and radiation curing.

Physical drying

The transition from liquid to a solid state by evaporation of solvents, including water, is called physical drying. The evaporation depends on the temperature, the air velocity and the concentration of solvent or water in the atmosphere. The higher the temperature the faster the drying time. Air circulation is also important for evaporating solvents or water. Even at a low temperature, but with a strong air blowing velocity a physical drying binder can be dried in a very short time. The third drying factor for physical drying is solvent or water concentration in the atmosphere. When the atmosphere is saturated with solvent or water, the new solvent or water from wet paint film cannot evaporate. Water in the atmosphere is called the humidity level. When it is very foggy, water-based physical dried paint cannot dry.

Examples of physical drying paints are nitrocellulose products or acrylic solution paints. After evaporation of the solvents and often water the coating is fully dry. White spirit in an alkyd-based paint also evaporates and enables physical drying of the product. After white spirit has evaporated it leaves a very sticky film, after which oxidative drying starts. The same situation occurs with two-component products and solvents. Solvents and water need to evaporate first before the chemical reaction can start.

Coalesce drying

All dispersion paints contain particles that need to flow together after the water has evaporated. This flow is called "coalesce". All

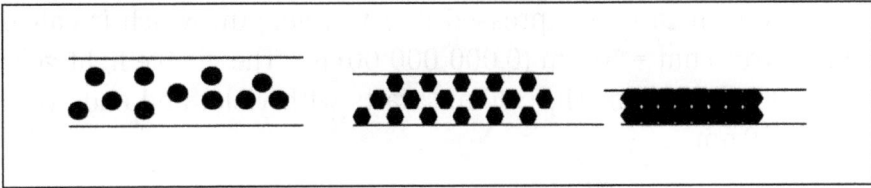

Figure 5.1: Three phases of the coalesce drying of dispersion paint

polymer particles must flow to form a homogeneous film. This migration process is called the "interpenetration" with which a co-solvent is very important. Co-solvents are solvents that are used to swell the polymer particles and evaporate very slowly. Well-known co-solvents are glycols, glycolesters (butylglycolacethate) and larger molecules of diolen.

Examples of coalesce drying dispersion paints are acrylic, polyvinyl acetate and polyurethane. In Chapter 3.1.2 the different finenesses of the dispersions have been described.

Chemical drying

During chemical drying the solid state is accompanied by an increase in molecular mass. The chemical reaction is dependent on the temperature and the amount of the reactance. If a two-component product is used and the reaction is incomplete, the coating is will not be completely dry causing insufficient coating properties. In addition, with radiation drying, all components need to be very precisely attuned. See Chapter 3.1.

5.2   Colour

Colours evolve from light. If we look at a spectrum, in which the light is broken, all the colours will appear in the same sequence: violet, blue, green, yellow, orange and red – a rainbow. The colour of an object is caused by a selective absorption of specific light frequencies from the atoms of the objects' surface. Red, green and blue lights are mixed "additives" to white light. When white light drops onto a surface and blue light is absorbed, the surface reflects red and green light – what we see as a yellow outlook. Pigments are chemical connections which can absorb one specific part of the spectrum.

The spectrum can be expressed in a wavelength, which is called nanometre. $1\,nm = 10^{-9}\,m$ (0.000 000 001 m). The visible light goes from 380 to 750 nm. The 380 nm starts with violet and ends with red at 750 nm.

In nature many colours adorn flowers and animals. People are able to see colours through the specific cones and bars in the retina in the eyes. Seeing colours also occurs through the blue, green and red sensitive nerve cells. Characteristic for colour seeing is tri-chromation that each colour can be mixed by variations in these tri-primary colour ratios. No two people look at colours in the same way. Small deviations in colour-seeing can be tested using the Ichihara pictures, in which several little coloured dots form figures. The deviations vary worldwide. In Europe and in the United States one out of twelve people have a colour disorder. In Asia one out of twenty people have a colour disorder and in central Africa, South America and Indonesia this ratio is one out of fifty people.

In each culture different colours represent various senses. History, religion and customs play an important part in understanding these senses. It is also interesting to learn about the various colour senses in the history of colours.

At present, colours can be measured by photoelectric cells. Each colour can be measured in colour tones (hue), saturation and in brightness. The hues are all the colours in the rainbow. The saturation is the proportion of colour within combined colours. An increased amount of one hue gives a higher reading than a lower amount. The colour is greyer. The brightness is the reflection value of a colour. White and yellow have a high brightness. Using these three values each colour can be recorded. Knowledge about colours can present a theory (apprenticeship) regarding the harmony of colours, colour relations, single or double colours, ton sur ton, etc. See also Chapter 7.3.

5.3   Protection

Substrate protection is a very broad subject, because different properties are allotted to each substrate. Some substrates are hard and durable (metal), other substrates are weak and weather sensitive (wood). In all fairness, protection is really what people want. The requirements for inside wall paints are totally different to the requirements for exterior wall paints. Car refinishing products also require totally different products to consumer coatings, etc.

The most protection properties are **resistance properties** against water, chemicals, temperature, scratches, wear and tear, fouling, and rubbing. In short: resistant to outside influences. The chosen binder in the product concerned is determined for each of the specific properties. The water resistance of an alkyd-based product is good for short term. A polyurethane product is much better, but neither of the products are suitable for under water application. Only epoxy and bitumen products can be applied under water – such as on the hubs of ships and yachts.

Different criteria determine the application range of coatings:

- Technological requirements, e.g. corrosion protection or furniture equipment
- Optical requirements, e.g. cars or doors of houses/buildings
- Number of objects to be coated, e.g. accessories or toys
- Size of work pieces, e.g. ships or bridges
- Special requirements, e.g. insulating wire or coatings for operating theatres
- Environmental requirements, e.g. solvent free coatings or lightweight metals

In practice, various requirements may arise simultaneously e.g. technical requirements (corrosion protection), size of work pieces (ships) and environmental requirements (without heavy metals).

5.4   Durability and sustainability

Durability is a complex topic. There are durable raw materials such as binders and pigments, which have a lasting effect. However, there are also durable raw materials, which are available for long-term lasting effects which is called "sustainability". Linseed oil is sustainable because there is a new linseed harvest each year. This is in contrast to acrylic binders, which are made from crude oil. The durability results are a long lasting protection period against influences that destroy the coating film. The protection of a coating film depends on the requirements. Each customer has his/her own specific requirements.

A coating durability is registered in years of protection properties, such as gloss preservation, no cracking, no discoloration, no chalking, no blistering and no erosion. The protection of the substrate has to be optimal, so that no corrosion, no rotting wood and no moist walls develop after the substrates have been coated.

The thickness of the layer is very important to several durability properties. Most paint products are applied in a two- or three-step system. The first step is to apply adhesion and filling, for which special primers or prime coats are necessary. The surface of the substrate must be sufficiently solid to provide a good base for a coating. A loose surface will not provide a good anchorage for a coating. Cleaning and sandpapering are necessary in order to prepare the surface for the paint application. Very open substrates such as plaster or plasterboard need to be sealed with sealer (insulating varnish). The second step is to build up the layer thickness with the selected paint. A good anti corrosion system has a thickness of 200 or 300 micron. The third step is finishing. A smooth flowing and firm paint surface is necessary so that pollution is avoided and it is possible to clean the paint surface.

An additional and important property is flexibility – the capacity of the coating to flow with the movements of the substrate. A wood coating has to be flexible to so that it adheres during expansion owing to temperature and moisture differences in the wood. Soft wood has a large extension coefficient. Similar properties are important and apply also to metal coatings.

Sustainable means that raw materials are available also in future. At the end of it all, our lifestyle, our manufacture, our energy must be sustainable. Raw materials such as extenders, pigments and binders are very important to the paint industry. Re-usable, recyclable or bio-based substances have become more important over the past five years. Undivided attention has recently been devoted to various projects such as 'from the cradle to the grave', the life cycle analysis, or from cradle to cradle with re-usable substances. A life cycle analysis is a critical view of all of the steps regarding the manufacture of raw materials, product composition and product use and to environmental effects such as energy use and recycling of all substances.

6 Paint products and paint formulas

Paint entrepreneurs are very reserved about their paint formulas. However, general characterisations can be deducted with regard to the composition of high gloss and mat paints, wood primers and metal primers, interior and exterior wall paint, plaster and wood stains. Many possibilities are available because of the implementation of solvent-based and water-based technology in all paint product groups.

Knowledge regarding paint formulas provides insight into the true properties so that when applied into practice, paint products can be easily recognized. Example: zone-marking paints on roads and on parking lots, without which there would be huge traffic problems causing impossible situations and parking disorders. The coatings that light up in the dark have also become indispensable to night safety on the roads. Having knowledge on the composition and properties of these types of coatings is invaluable.

In addition, understanding of technical "product information sheets" is much easier when one has background knowledge regarding paint formulas. In this chapter, paint formula principles, paint product principles and several paint product formulas will be discussed. The paint formulas described in this chapter are only rough indications used for illustration and better understanding of paint formulas and properties.

6.1 Paint formula principles

The basis of each paint formula is the binder. The binder type and the amount of binder used in a paint formulation decide, for a large part, the paint properties. In Chapter 2, paint ingredients and binder properties were discussed. Each binder is a compilation in a special

Adrie Winkelaar: Coatings Basics
© Copyright 2009 by Vincentz Network, Hannover, Germany
ISBN 978-3-86630-851-0

Table 6.1: Examples of formula principles

Paint formulas	Formula 1 strong coating	Formula 2 low solid impregnating	Formula 3 porous coating	Formula 4 low solid colouring
Binder (solid)	50	25	10	5
Pigments/ extender	20	5	60	25
Additives/ solvent	30	70	30	70
Total weight %	100	100	100	100
Solid content	71 %	31 %	71 %	31 %
Pigment- binder ratio	1 : 2,5	1 : 5	6 : 1	5 : 1
PVC	14	7	71	66

binder form – a solution, dispersion or an emulsion (see Chapter 3). With regard to dried paint properties, this special formula is not so relevant. The binder pigment ratio is very important in addition to the solid content of a formula. Each formula has an ingredient amount of 100-weight percentage. In Table 6.1, four formulas with different binder percentage of 50, 25, 10 and 5 % are mentioned. The solid content of each formula is calculated and the pigment binder ratio is reproduced.

The pigment volume concentration (PVC) is the ratio of the pigment volume to the total volume of the solid coating. In this pigment volume all dry pigments and extenders have been added. The volume of the dried film is the sum of all solid binders, resins, plasticisers and additional substances with a solid content.

The formula is

$$PVC = \frac{\text{Volume Pigments and Extenders}}{\text{Volume Pigments and Extenders} + \text{Volume Binders and resins}} \times 100$$

For calculation of the pigment and binder volume it is necessary to know the densities of the pigments and binders. Generally speaking, the solid binders have a density of $1.2\,g/cm^3$. The pigments and extenders have different densities, for example: titanium dioxide = $4.1\,g/cm^3$, carbon black = $1.8\,g/cm^3$, phtalocyanine blue = $1.6\,g/cm^3$, calcium carbonate = $2.8\,g/cm^3$ and iron oxide red = $5.0\,g/cm^3$

A strong coating (Formula 1 in Table 6.1), has a high binder percentage because properties, such as, hardness, flexibility and gloss are dependent on the binder amount. This paint is called an enamel or high gloss coating. The second formula is a low viscous impregnating product for wood or plaster. It is a low solid with low PVC, which means that there is enough binder in the paint to apply to a wooden or plaster substrate.

The third formula, a highly filled porous coating, has very high PVC, which means that there are large volumes of pigments and extenders for hiding power and filling properties. The extender amount is also responsible for vapour absorption when applied on the interiors of houses. These properties are important especially when being used on ceilings and walls. The final formula (number four in Table 6.1), is an artistic paint for interior colour substrates. The low binder percentage offers no protection but it is sufficient for binding between the pigments. Many solvents (water) provide good surface application and good colour distribution. Dried film has poor powder properties (caulking), but after painting a clear varnish can be applied for protection.

6.2 Paint product principles

Each paint formula is compromised in price, properties and application. For example, if the user wishes to paint a main entrance door to a property, a strong coating and an attractive finish (formula 1 in Table 6.1) would be needed. And if a user wishes to paint a bedroom wall or a ceiling, then an inexpensive paint with good hiding power properties would be adequate for the job (formula 3). For this reason, there are many paint products available, each with different properties and having different prices.

The primary paint properties are those that adjust to specific painted substrates. Wood is a natural substance; however, hard wood and soft wood have different properties. On soft wood the first layer has to penetrate to enable good adhesion. For adhesion on metal it is necessary to use a special metal primer. The same applies to stones or concrete those also have other properties.

All paint products can, therefore, be divided into product groups, such as, products for wood, metal, stone/concrete and other substrates such as, plastics, laminates and sheet plates.

6.2.1 Wood

The water absorption of hard or soft wood varies. Therefore, soft wood has to be protected more than hard wood, with more layer thicknesses again moisture and humidity. The application of primers and different layers are necessary for the protection of soft wood against wood rot. Coating flexibility is also necessary for protecting against the expansion of soft wood during temperature changes. Alkyd and acrylic binders are usually applied in wood protection coatings.

In addition, wood primers, lacquers and wood stains are available for the protection of wood against weather influences. The wood stain coating that is used to impregnate soft wood will erode after some years; however recoating is very easy to carry out. Large-scale wood staining to houses, especially farmhouses is very attractive, because the natural wood structure is enhanced. There are varnishes and translucent wood stains available and because they are very sensitive tor UV light, they have short exterior durability – however UV absorbers are used to extend the durability of varnishes and translucent wood stains.

6.2.2 Metal

Iron, copper, zinc and aluminium are different metals with different properties. Weather influences (oxidation) also have different results on each metal so that conductivity and metal expansion are important properties for recoating. An epoxy primer is usually used because good adhesion is important. A watertight coating is relevant for the prevention of metal oxidation. In practice, the thickness of the layer proves to be the best solution. More than a 200 micron layer thickness is usual for exterior application and more than a 300 micron for under water applications. Product names for building up the layer thickness are "body coats" and "surfacers".

Special anti-corrosive primers are used on iron/steel surfaces. There is an anti-fouling paint for under water usage and for the interiors of drink water tanks there are special toxic free epoxy coatings available.

6.2.3 Stone/concrete

Bricks, concrete and plaster are different substrates for coating applications. Important properties are the water absorption, the porosity, the alkalinity and the hardness or softness of the substrate. Very porous ceilings or walls, treated with plaster, need a sealer application, after which an open wall paint containing a high PVC value for the absorption of water vapour outside of the room is usually applied. A concrete substrate is very hard and less porous. For exterior application, an adhesion primer is necessary, after which a coating can be applied. Between these two extreme possibilities, various acrylic dispersion binder formula wall paints can be applied. For special situations, there are water resistant coatings based on a two-component emulsion epoxy, available.

6.2.4 Other substrates

Plastics and laminate sheets are increasingly being used. In addition, old coating systems can be assessed as laminate sheets. The most important focus point is to achieve a good adhesion. Cleaning and sandpapering are the most usual methods.

The secondary properties of paint products are dependent on the demands of the user or customer. A cheap product contains less binder, less pigments and more extender. The special binder properties are usually less, because extenders replace them. For this reason a huge range of formulations are possible – in practise no two paint products have the same properties.

Table 6.2: Cheap wall paint formulas

	Raw material price Euro/kg	Cheap formula	Very cheap formula
Acrylic dispersion (50 % in water)	4.00	20	10
Titanium dioxide	5.00	10	5
Extender	0.25	40	55
Additives (thickener, dispersion agents, antifoam and biocides	5.00	4	4
Water	0.00	26	26
Total in weight %		100	100
Raw material price Euro/kg		1.60	0.99

In table 6.2 an average raw material price is given for the main ingredients (column 2). In column 3 and 4 two cheap wall paints are mentioned. To decrease the binder and pigment percentages (in weight) the raw material price is always lower. A high titanium dioxide percentage offers a better hiding power, but is more expensive.

6.3　　Paint product formulas

In the paint market there are differences between decorative paints. The professional quality and the do-it-yourself quality are usually equal. Industrial products are made-to-measure products that are applied and dried in a different way to the decorative products. The decorative products are applied by brush, roller or spray and are dried under normal conditions and all these products have to follow the European Directive 2004/42/EC (see Chapter 10).

The follow formulas are examples and indications of some of the many paint products. They provide an insight to the different products and qualities.

6.3.1 Products for wood

Wood products can be found in several market areas. Window frames and doors in houses, wooden constructions and wooden panelling, wooden floors (parquet) and wooden tables, chairs, etc. In gardens there are also wooden borders, garden houses, and garden furniture etc.

Wood primer for interior and exterior application needs filling properties and must be sandpapered. The solid content is 80 % (high solid) and the PVC is 54. The solvent can be white spirit, but water is also possible if the alkyd is emulsified.

Table 6.3: Wood primer formula for interior and exterior

Raw material	Parts per weight
Linseed oil alkyd (100%)	20
Titanium dioxide	10
Extender (calc and talc)	47
Additives (dispersion agent, drier and anti peeling)	3
Solvent	20
Sum	100

High gloss enamel contains no extenders, since they would reduce the binder gloss. In silky gloss and mat enamels, flatting agents are added – such as very fine extenders or crystal silica; the solid content must be high. The solid content is 80 % and the PVC is 15.

Table 6.4: Formula of a high gloss enamel

Raw material	Parts per weight
Soya alkyd (100 %)	40
Titanium dioxide	30
Additives (dispersion agent, drier and anti peeling)	5
Solvent	25
Sum	100

Opaque wood stain for exterior application contains less filling properties as a wood primer. The extenders are porous and are good vapour permeable. More binder is necessary for good flexibility. The solid content is 57 % and the PVC is 29.

Table 6.5: Formula of an opaque wood stain for exterior application

Raw material	Parts per weight
Alkyd emulsion (45 % in water)	60
Titanium dioxide	15
Extender (talc and clay)	15
Additives (dispersion agent, drier and anti peeling)	4
Water	6
Sum	100

Translucent wood stain contains a small amount of translucent pigment, such as, transparent micronized iron oxides. If flatting agents are used, extenders and crystal silica are not used because the white haze they produce. In this case synthetic waxes are used, such as polyethylene or polypropylene that provide a low gloss finish on the top of the translucent film. The solid content is 42 % and the PVC is 13.

Table 6.6: Formula of a translucent wood stain coating

Raw material	Parts per weight
Alkyd emulsion (45 % in water)	70
Transparent iron oxide	5
Polyethylene	5
Additives (dispersion agent, drier and anti peeling)	5
Water	15
Sum	100

Parquet varnishes are applications for the decorative markets and the industrial markets and need pre-treatment. Polyurethane binders are used on a large scale. The PU binder is expensive and for this reason combinations of cheaper acrylic binders are used. The flatting agent is also a polyethylene substance used for application on dark wood. The solid content is 35%; the PVC is very low because there are no pigments, only flatting agents.

Table 6.7: Formula of a parquet varnish

Raw material	Parts per weight
Polyurethane dispersion (50 % in water)	30
Acrylic dispersion (50 % in water)	30
Flatting agent	5
Additives (anti foam and thickener)	3
Water	32
Sum	100

An UV furniture drying varnish is clear and dries in one second under a UV light. These types of formulas contain a UV initiator and reactive binder such as "polyester acrylate" and a reactive substance, for example, reactive solvent. The solid content is 86%.

Table 6.8: Formula of a UV furniture varnish

Raw material	Parts per weight
Polyester acrylate (100 %)	75
Reactive solvent	8
Silica gel (20 %)	8
Flatting agent	4
Additives	3
UV initiator	2
Sum	100

6.3.2 Products for metal

For these types of products there are three different market sections which distinguished as follows: the professional steel conservation (1) such as ships, bridges and constructions; the decorative market (2) for professional painters and do-it-yourself applications and the industrial steel market (3) for metal products and components for industrial areas.

A two component anti-corrosion primer is a commonly used product. Increasingly more water-based products are being introduced in the market. An epoxy primer is best for adhesion and together with anti-corrosion pigment; an anti-corrosion primer can be made. The solid content is 48 % and the PVC is 24.

Table 6.9: Formula of a two component anti-corrosion epoxy primer

Raw material	Parts per weight
Component 1	
Epoxy emulsion (55 % in water)	50
Iron oxide red	8
Extender (talc)	12
Zinc phosphate	10
Additives (dispersion agents, anti foam)	2
Water	18
Sum	100
Component 2	
Amine adduct	10

Two-component enamel is usually made with polyurethane or epoxy binders. The epoxy is sensitive to UV light and will become mat in time. With the use of a polyurethane binder, high gloss two-component enamel can be produced. Increasingly more solvent-free coatings are being developed.

Table 6.10: Formula of a two component polyurethane primer

Raw material	Parts per weight
Component 1	
Polyol	38
Additives (water absorption agent, dispersion agent, anti foam)	8
Extender	30
Pigments	5
Sum	81
Component 2	
Polyisocyanate	19

A standard anti-corrosive primer for do-it-yourself applications can be made using an alkyd binder and lead-free pigments. The solvent is usually white spirit, which is replaced with water. Below is a classic formula with solid content of 70 % and PVC of 40.

Table 6.11: Formula of a standard anti-corrosive primer for do-it-yourself applications

Raw material	Parts per weight
Linseed oil alkyd (80% in white spirit)	30
Iron oxide red	15
Extenders (barium sulphate and talc)	20
Zinc phosphate	10
Additives (dispersion agent, drier and anti peeling)	5
Solvent	20
Sum	100

Industrial stoving enamel is one component within two components that react at high temperatures. Alkyd reacts with melamine resin at 120 to 160 °C. The solid content is 75 % and the PVC is 16

Table 6.12: Formula of an one component industrial stoving enamel

Raw material	Parts per weight
Alkyd (75 % in butyl glycol)	40
Melamine resin (80 % in butanol)	10
Titanium dioxide	20
Additives (dispersion agent, amine for neutralizing and catalyst)	4
Water	26
Sum	100

Powder coatings are being increasingly developed. These products are manufactured in paint factories, after which, the coating is broken up and milled to powder. The powder can be applied through an electrostatic spray onto substances that are connected with an earthed wire. Afterwards the coating needs to be stoved at a high temperature. The solid content is about 100 % and the PVC is 12.

Table 6.13: Formula of powder coatings

Raw material	Parts per weight
Epoxy resin (100 %)	45
Phenol crosslink	25
Iron oxide red	12
Extenders (barium sulphate and talc)	15
Additives	3
Sum	100

Filler or putty is necessary for repairing cars and/trucks. A putty knife is used for the application or a spray. There are one and two component fillers. The most famous filler is based on polyester. The solid content is 95 % and the PVC is 56.

Table 6.14: Formula of a two-component automotive or trucks repair filler

Raw material	Parts per weight
Component 1	
Polyester (65 % in styrene)	30
Titanium dioxide	5
Extenders (chalk and talc)	60
Additives (plasticizer)	5
Sum	100
Component 2	
Peroxide (50 % in plasticizer)	3

6.3.3 Products for concrete and stone

This is the biggest product group in the paint industry: the application of large surfaces in buildings using interior and exterior wall paint, paint for ceilings, concrete floors and concrete surfaces on bridges, flyovers and viaducts. Stones, plasters and concrete are important surfaces for colouring. Since houses and offices are used intensively, wall paint protection is also important. All large surface products can be thinned with water, especially those based on acrylate dispersions

Indoor dispersion wall paint is a very cheap product that comprises a lot of extenders for absorption in the vapour content. Because the PVC is very high, the wet-rub resistance is very important to distinguish the different wall paint qualities. The solid content is 65 % and the PVC is 83.

Table 6.15: Formula of an indoor acrylate dispersion wall paint

Raw material	Parts per weight
Acrylate dispersion (50 % in water)	10
Titanium dioxide	5
Extender (chalk and calcite)	55
Additives (dispersion agents, anti foam, thickener and biocides)	4
Water	26
Sum	100

Outdoor dispersion wall paint is more expensive because more binder is used. Also finer and better resistance extenders are used. The solid content is 58 % and the PVC is 46.

Table 6.16: Formula of an outdoor acrylate dispersion wall paint

Raw material	Parts per weight
Acrylate dispersion (50 % in water)	35
Titanium dioxide	15
Extender (calcite and talc)	25
Additives (dispersion agents, anti-foam, thickener and biocides)	5
Water	20
Sum	100

Indoor decorative plaster needs to provide a thick layer on walls. This means that more coarse ingredients are necessary. Different sand fractions are used to obtain a good ball-pilling up. For this reason no cracks will appear in the thick layers after drying. The solid content is 78 % and the PVC is 75.

Table 6.17: Formula of an indoor decorative plaster

Raw material	Parts per weight
Acrylate dispersion (50 % in water)	20
Titanium dioxide	5
Extender (chalk and calcite)	30
Sand fractions	30
Additives (dispersion agents, anti-foam, thickener and biocides)	5
Water	10
Sum	100

Hard silicate wall paint is possible in contrast with thermoplastic acrylate. Silicates react with water and carbon dioxide from the atmosphere to become hard polymer (silicification). The microstructure of silicate coatings provides good capillary water absorption properties. The solid content is 50 % and the PVC is 60.

Table 6.18: Formula of a hard silicate wall paint

Raw material	Parts per weight
Silicate solution (30 % in water)	35
Titanium dioxide	10
Extender (chalk and calcite)	30
Additives (dispersion agents, anti-foam, stabilizer)	4
Water	21
Sum	100

A concrete floor coating can be made with a combination of polyurethane and acrylate dispersions. PU has good abrasion resistance and is also resistant against chemicals. The combination with the acrylate dispersion is for the raw material price reduction only. The solid content is 55 % and the PVC is 40.

Table 6.19: Formula of a concrete floor coating

Raw material	Parts per weight
Acrylate dispersion (50 % in water)	25
Polyurethane dispersion (50 % in water)	25
Titanium dioxide	10
Extender (chalk and calcite)	25
Additives (dispersion agents, antifoam, thickener, biocides and anti-slip agents)	5
Water	10
Sum	100

The majority of floor coatings and systems are two-component products. In addition, water based products and solvent-free products appear in this market. A two-component epoxy flow floor compound is filled with different sand fractions such as the above-mentioned decorative plaster, to obtain good ball pilling up. When the ball pilling up is not in balance, cracks in the thick layers may appear after drying. The solid content is about 100 % and the PVC is 24.

Table 6.20: Formula of a two-component floor coating

Raw material	Parts per weight
Component 1	
Epoxy resin (100 %)	40
Pigments (titanium an iron dioxide yellow)	5
Extenders (calcite and talc)	15
Sand fractions (0.1 to 0.5 up to 1.0 mm)	35
Additives (dispersion agents, viscosity agents)	5
Sum	100
Component 2	
Polyamide (100 %)	20

6.3.4　Other products

Besides the above mentioned product groups, there are many special technique products available in the paint market. Repair fillers, special plastic primers, aluminium primers, caulking compounds and zone-marking paints have been developed for various consumer and industrial markets.

Plastics and different laminate sheets appear in all market sections. In the decorative market, the automotive market (new cars and trucks) and industrial market (consumer products) more and more plastics are being introduced. Not all of these plastics can be recoated in the normal way. Hard plastic materials can be recoated, but soft plastic materials need special pre-treatment.

A decorative plastic primer has to have good adhesion, in addition to good flowing or levelling properties. The solid content is not so high (40 %) and the PVC is 35.

Table 6.21: Formula of a decorative plastic primer with good adhesion properties

Raw material	Parts per weight
Acrylate dispersion (50 % in water)	30
Titanium dioxide	10
Extenders (calcite and talc)	15
Additives (dispersion agents, levelling agents, biocides)	5
Solvent (somewhat solving substrates)	5
Water	35
Sum	100

Instant wall filler is ready mixed filler, which, compared to the old fashion gypsum, is mixed with water before use. Instant filler has to be applied with a filling knife and dried filler must be able to be sanded and recoated with normal wall paint. The shrinkage of the instant filler depends on the amount of water. The solid content has to be fixed as high as possible (75 %) and the PVC is 75.

Table 6.22: Formula of an instant wall filler (ready mixed)

Raw material	Parts per weight
Acrylate dispersion (50 % in water)	20
Extenders (chalk, calcite and talc)	70
Additives (dispersion agents, viscosity agents,...)	5
Water	5
Sum	100

Zone-marking paints are available in two qualities: solid thermoplastics, which are applied in either a hot liquid consistency and/or a liquid quality that is based on an acrylate solution or dispersion. After application, glassy balls have to be sprinkled onto the wet paint, so that white stripes will lighten up in dark surroundings. A liquid zone-marking paint based on acrylate dispersion is almost similar to concrete floor paint. The solid content is 62.5 % and the PVC is 50

Table 6.23: Formula of a liquid zone-marking paint based on acrylate dispersion

Raw material	Parts per weight
Acrylate dispersion (50 % in water)	35
Titanium dioxide	15
Extenders (calcite and talc)	30
Additives (dispersion agents, antifoam, viscosity agents)	5
Solvent (to pull up the asphalt)	5
Water	10
Sum	100

7 *Paint production*

Paint manufacturing comprises mixig, stirring and distributing pigments, extenders and flatting agents into binders. Each paint formula process begins with dispersion, which is self-manufactured or purchased. There is a huge increase of specialised companies that concentrate on manufacturing pigment dispersions. Since the colour mixing machines offer a vast range of colour making possibilities, both in own manufacturing (ready mixed colours) as well as in the market store houses and shops, there is an enormous demand for pigment pastes.

Paint production processing depends on the size of the manufacturer. Small quantities of between 500 or 1000 litres are produced in movable tanks and large quantities of between 5,000 or 10,000 litres are produced in fixed installations with pipes and pumps attached.

Paint formulas are developed in the manufacturers own laboratory after which the manufacturing process is carried out step by step. The quality control department inspects each batch of manufactured paint to ensure they consist of the correct properties. After filtering, the paint is dispensed into tins or buckets. The logistics process in paint factories, as well as paint markets are both very complex – these departments have to make sure that the correct colour and correct quality is delivered to paint users. This chapter describes the production process and the typical phases of grinding, mixing, filtration and filling tins.

7.1 *Production process*

The paint production processes can be divided into continuous processes and batch processes. Each process comprises many sub-processes for the manufacturing of pigment dispersions, resins solutions, flatting agent dispersions, additive solutions, etc. A paint formula con-

Adrie Winkelaar: Coatings Basics
© Copyright 2009 by Vincentz Network, Hannover, Germany
ISBN 978-3-86630-851-0

tains a large number of raw materials that require a high standard of operating, cleaning and maintenance. All paint factories have a huge range of finished products and the administration, inspection, storage and distribution costs were very high. Over the past thirty years many changes have taken place, with the introduction of automation – computer driven processing and automatic manufacturing processes that have resulted in costs being reduced. At the same time, numerous safety and environmental regulations relating to raw materials, production processes and final products have been implemented.

7.1.1 Raw materials

Paint makers purchase their raw materials from the chemical industry that manufactures pigments, binders, additives and solvents. Information regarding paint formulas has been mentioned previously in Chapter 6. In 2005, the average national weight of paint formula in The Netherlands was 30-weight% binder, 30-weight% pigments and extenders, 4-weight% additives, 20-weight% solvents and 16-weight% water. Increasingly more water-based products are

Figure 7.1: 200 litre vessels used for binders and solvents

Figure 7.2: 25 litre buckets used for additives

Figure 7.3: 25 kg bags used for pigments and extenders

Figure 7.4: 20,000 litre tank storage for binders

being introduced and there is a slow increase of high solid products. This means that a large amount of binders, pigments and extenders is necessary for standard manufacturing. Tanks (10 or 20 tonnes), IBC (1000 litres), vessels (200 litres) and bags (25 kg) are the packaging mostly used for raw materials. They are the input of the manufacturing process.

Figure 7.5: Filling and weighting additives during manufacture

The weighing of all these substances is the most characteristic of paint manufacturing. Large volumes of solvents and binders are pumped through the volume measuring equipment through the pipes and into tanks. Powders and additives are weighed on scales inside movable vessels.

7.1.2 Processing

All raw materials are usually on stock in storage. The first step in the paint manufacturing process is the manufacture of dispersions of powders, such as pigments, extenders and flatting agents. Increasingly more colour tinting pastes are being used which are purchased from specialist suppliers. Step two sees the start with the placing of the binder into the production barrel or tank in which the sub-dispersions (step 1) are mixed. Step three is the mixing of all the additives together. Step four is mixing in the right colour and the testing of the quality of the batch. The final steps are filtering and filling containers (tins or buckets).

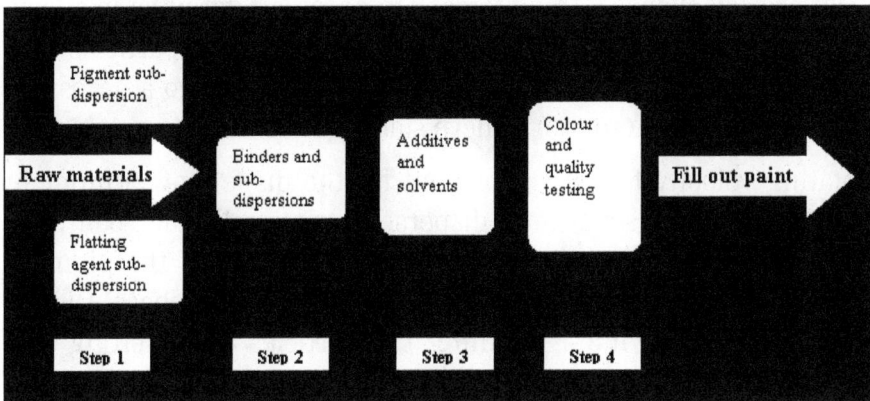

Figure 7.6: Production steps

The dispersion is the most complex stage of paint production. During the other steps all of the ingredients are mixed together. In general, all paints are produced using the same manufacturing process. For the production of varnish, the process starts at step 2. A number of batches of the same type are processed simultaneously, which enables a continual flow of production and quality.

7.1.3   Equipment

Mixers, barrels, pipes, pumps and scales are the most important equipment utensils in the paint industry. The mixing of substances is based on various physical and technological principles. The main tool for stirring and mixing is called a 'stirrer'. There are many stirrer types that provide different horizontal and vertical flowing within the stirring vessel. A propeller stirrer is used for vertical flow patterns (axial) and a disc stirrer is implemented for creating two flow patterns in one stirring vessel (radial) – one at the bottom and one at the top of the stirring vessel.

The mixing principle is a homogenising process. The viscosity is also important for the realization of a homogenising product. The final position of the stirrer in the stirring vessel is very important to the process.

7.2   Grinding and milling

All powder substances such as pigments, extenders and flatting agents have to be grinded into a fluid binder so that optimal properties in the paint are established. Strong forces are necessary for breaking down the powder agglomerates and for grinding into fine particles (see Chapter 3.3). Afterwards the primary particles undergo a necessary wetting process in order to achieve stable quality results.

Rotating velocity forces are present for the dispersing equipment such as agitators, stator-rotor dispersers, triple roll mills, ball mills and extruders. Dispersion means breaking-down of the pigment agglomerates that are finely distributed in the liquid phase. Dispersion is often known as "grinding" the particles and "wetting" the particles surfaces. Inorganic pigments have a particle size range of 0.1 to 10 micron and organic pigments have a particle size range of 0.05 to 10 micron.

7.2.1   Agitator or dissolver

An agitator or dissolver can be defined as a high-speed serrated-edge disc mill that is used for dispersing extenders and pigments in wall paints and primers. In addition, for the pre-mixing of pigments, the dissolver is the first step before milling it with other equipment.

Figure 7.7: Dispersing machinery

Figure 7.8: Dispersing machinery in action

Figure 7.9: Dispersing tank with high-speed
serrated-edge disk

The serrated edge on the disc
is oriented to the circum-
ference in such a way that
excess pressure is generated
on the outer perimeter of the
serrated-edge disc. Reduced
pressure is generated on the
inner side. By shearing, the
agglomerates break down
into smaller pieces, such as
primary particles or individual particles. Since the
serrated-edge disc results in a
dispersion effect, the peripheral velocity of the disc v_p has
to be regarded as the most
important value of the oper-

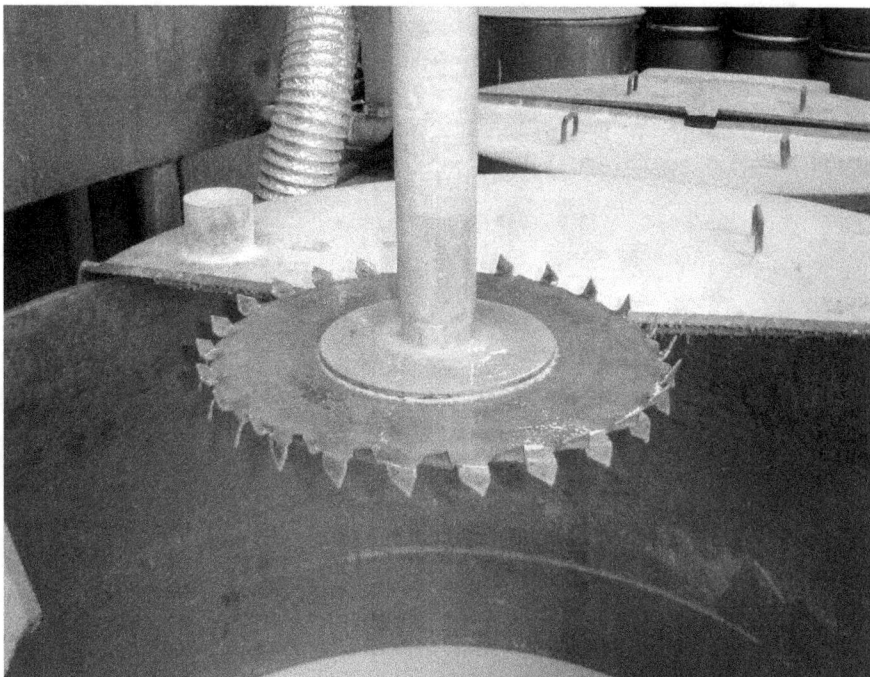

Figure 7.10: Close up of the high-speed serrated disk

ating status of the dissolver. This is calculated from the speed n and the disc diameter D with π (= 3.14) as a constant

$$V_p = \pi \cdot n \cdot D$$

Typical peripheral velocities are, for example, 10 m/s for water based paints and 25 m/s for solvent based paints. The average particle sizes that can be obtained are approximately 10 micron or higher, which is insufficient for topcoat applications.

The correct flow pattern in the dissolver vessel is very important. The material to be dispersed must be poured in a spiral pattern from the sides towards the disc so that all of the particles included. To establish this flow pattern, the disc distance to the vessel bottom is 0.5 to 1.0 D (the disc diameter). The total volume of the vessel is 2D high. The viscosity also needs to be set to an optimal level. Viscosity that is too low does not provide a good shear flow and viscosity that is too high causes problems with the equipment and leads to high temperatures. Modern dissolvers automatically adjust their speed to the changing viscosity during the dispersion process. If, after 10 minutes the temperature increases, the viscosity decreases. The average dispersion time of a dissolver is approximately 20 to 30 minutes. Some dissolvers are equipped with wall scrapers that scrape the particles from the wall and back into the flow pattern in the vessel.

7.2.2 Stator-rotor dispersers or attrition mills

Stator-rotor disperser equipment comprises of balls that act as grinding media. The dispersing effect is caused by the balls operating together in translational and rotational movements, the result of which is an impact caused by both balls colliding together and against the wall in the grinding compartment. In the

Figure 7.11: Horizontal pearl mill disperser, source: Bühler, Switzerland

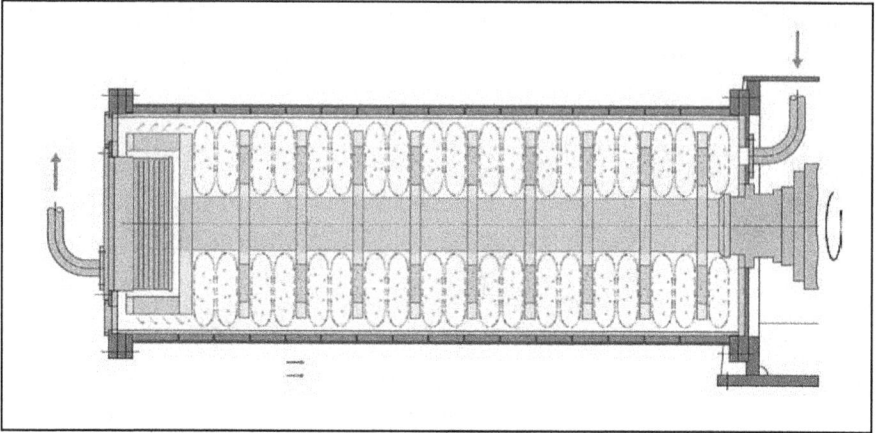

Figure 7.12: Cross section of the horizontal pearl disperser, source: Bühler, Switzerland

grinding compartment a stirrer or rotor with discs moves slowly through the pigment dispersion, which, at the same time, is being pumped through the compartment. There is vertical open equipment (sand mill) and small sand-like balls. There are closed vertical attrition mills and closed horizontal attrition mills.

This stator-rotor equipment comprises a filling mass of fine balls (beads) with a gross capacity of approximately 250 litres. The beads

Figure 7.13: Vertical perl mill disperser, source: Bühler, Switzerland

are made from materials such as steel, zirconium oxide, aluminium oxide, silicium oxide or a combination of these oxides. Glass, steatite (modification of talc) and plastics are also used. The diameter lies in the range of 0.1 to 3 mm. The harder the beads are, the bigger the dispersion intensity is which, at the same time, increases the wear and tear of the mill. The density of the beads influences the dispersion results. The density of the beads, the pigment dispersion viscosity and the type of beads and diameter are all-important for an optimal grinding process. Steel beads cannot be used for dispersing light coloured pigments, because of the dark grinding dust they produce.

The attrition mill is a continuous circulating process. The pigment dispersion is conveyed through the mill at various levels of excess pressure up to a maximum of 6 bar. A short residence time leads to poor dispersion and

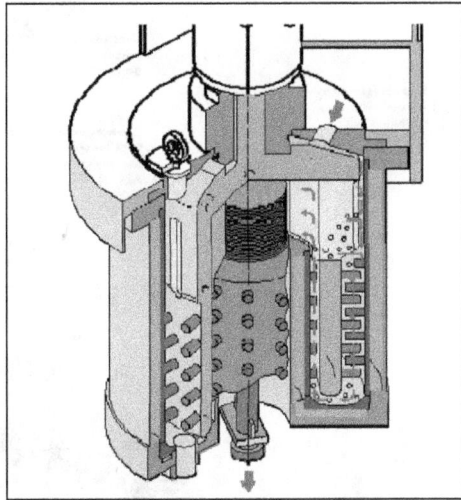

Figure 7.14: Cross section of the vertical perl disperser, source: Bühler, Switzerland

a long residence time will result in good dispersion. While the residence time distribution can only be determined by experiments, the usual residence time t can be calculated from the material throughput V_m (volume flow rate) and the free grinding compartment volume V_g.

$$t = \frac{V_g}{V_m}$$

The free grinding compartment volume is equal to the total grinding compartment volume minus the total volume of the beads, the rotor and the discs. In practice this means that the residence time is approximately just a few minutes.

7.2.3   Triple roll mill

Triple-roll mills are rarely found in the paint industry because they have been replaced by stator-rotor dispersers and colloid mills. However, triple-roll mills are still used to grind pigments in the printing ink industry. The basic functioning of a triple-roll mill is the different speed ranges of the three rolls. In the two 'nips' of the three rolls the shear flow is very intensive and provides the grinding effect. The first roll is the feed roll with a feed hopper, in which the pigment disper-

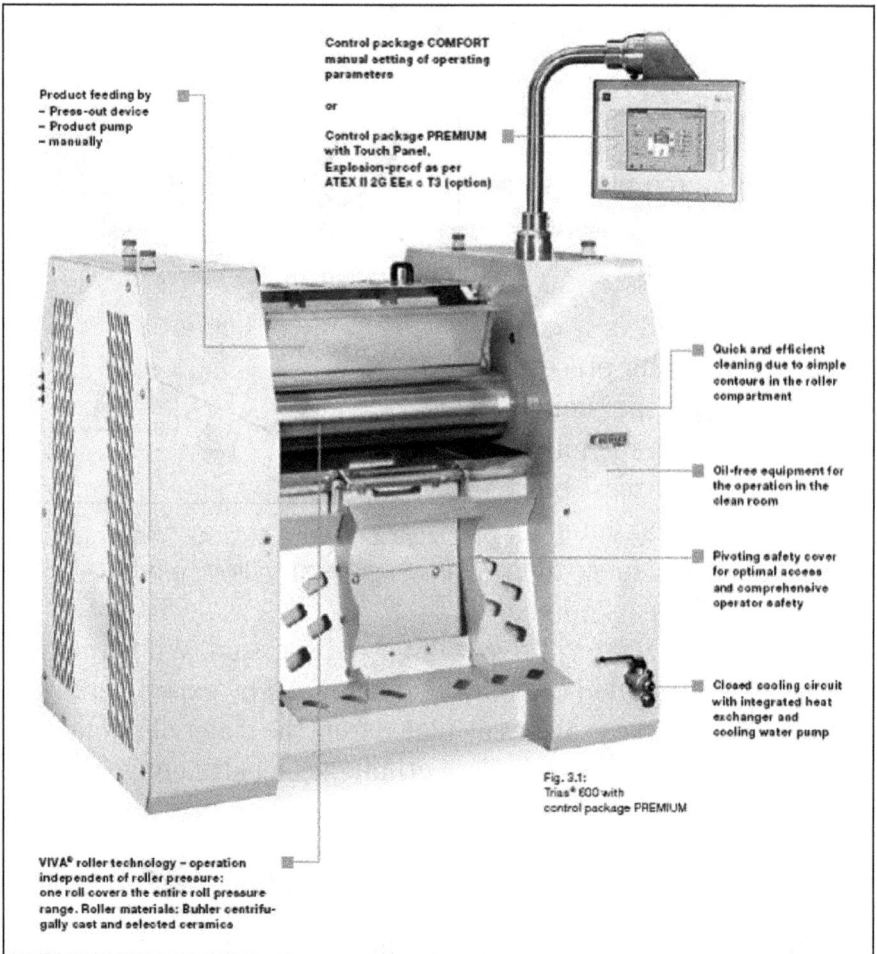

Figure 7.15: Triple roll mill, source: Bühler, Switzerland

sion is added. The third roll is the take-off roll, from which the pigment dispersion can be collected. Important is the contact pressure of the rolls and the rotating velocity (figure 7.15).

The viscosity of the pigment dispersion must be as high as possible to obtain high shear forces. From highly concentrated pigment, pasty materials are obtained. There is a relatively low throughput rate.

7.2.4 Extruder or kneader

The central component in the extruder is the screw, which rotates in a vessel or a cylindrical runner. The screw draws the substances in the vessel or the cylindrical runner for kneading. This equipment is used for plaster, fillers, putties and powder coating. Due to the high shear forces in the extruder, the pigment and extender agglomerates are broken down into a peeling action. A critical point is the temperature, which increases anyway from the kneading process.

For powder coating manufacturing, the melt in the extruder must be below the stove temperature of the powder coating. This means that the residence time should be no longer than absolutely necessary for the dispersion. A temperature range is 80 to 100°C and the main residence time is between 15 to 20 seconds.

7.3 Colour mixing

There are more than 25,000 colours used for paint application. To cater for the demand of all the colours, paint makers almost always use exclusive pigments that can be divided up as follows:

- yellow,
- orange,
- red,
- violet,
- blue,
- green,
- brown,
- black and
- white.

The average person can observe more than 7 million colours and distinguish more than 100,000 colours. In the paint industry the colour names that are used are based on the pigment composition e.g. titanium dioxide (white) and phtalocyanine (green). It is possible to acquire 25,000 colours with mixing, however it is impossible to give each colour a name. To avoid making mistakes, *A. H. Munsell* from the U.S.A. has developed a systematic, three-dimensional colour range:

- colour tone (hue),
- colour saturation (chroma) and
- colour lightness.

When colours are made, it is important to have knowledge of this theory, because the pigment suppliers provide information regarding the pigment **colour**, which is mixed, in several steps, with white. The **lightness** in the same colour tone is made higher and higher.

If the lightness is less, the colour tone is strong, but the difference between high or low colour tones is the **saturation**. Fierce red has a high saturation, but a somewhat greyish red has less saturation.

There are three colours, which cannot be made by mixing, they are called **primary colours** and are:

- yellow,
- red and
- blue.

Mixing two of the primary colours together will result in one of the **secondary colours**:

- orange,
- green and
- purple.

These six colours provide the **colour circle**:

red – orange – yellow – green – blue – purple

These six colours can be mixed to make twelve colours, the twelve resulting colours can be mixed to form twenty-four and so on – it is almost a never-ending story. The mix-ratio is also very important. A 50/50 mix of red and yellow would result in a colour that is totally different from a 10/90 mix.

Black and white are very important in colour mix formulas. To create colours that are greyer or lighter, black and white pigments can be added. A mixture of two complementary colours with the same ratio, such as red and green will also result in black. **Complementary colours** are the opposites of the six-colour circle.

7.4   Filtration and filling

Before becoming decanted paint, undesirable particles, e.g. dust, gel particles, oversized pigment or extender particles, must be removed by means of filtration. In this process two filtration types are distinguished: surface filtration and depth filtration.

Surface filters, which are also known as screens, consist of a screen cloth with a defined uniform mesh size. The uniform mesh size is available in various ranges, for example 5 to 800 micron. Particles that cannot pass through the mesh are separated off as screening residue.

Depth filters are porous plates, pipes or bags of varying thicknesses through which the paint is pumped. The characteristic value for depth filters is the "retention rate". This rate depends on the filter element and the physical filtration conditions, such as viscosity, pressure and temperature.

Figure 7.16: Filtration

Figure 7.17: Filling

The most frequently applied filter equipment in the paint industry is the bag filter, which can be moved to different parts of the manufacturing process. The filter bag is attached to a filter-house, through which the paint is pumped under pressure.

Filling out the paint into tins, containers or buckets is a labour-intensive activity. It is sometimes undertaken manually, but the majority of the paint industries are equipped with semi or fully automatic systems. Increasingly more pneumatic technology is being implemented for filling out into small tins from a storage tank. All empty tins are retrieved on a turn-around basis and are then dispatched to a production line, where they are filled and sealed with lids. They are then dispatched to another turn-around point for packaging in boxes. The boxes and/or buckets are stacked on pallets for storage and/or transportation.

8 Application, drying and removal

From an early age we have all had experience in applying paint. Schools include art and painting in their education curriculum and children are motivated to express their feelings with the use of a pen or brush and paints on paper. For the very first coloured sketches paints were nearly always used sketches.

Later on in life, people make another important step when they leave their parental home and buy or rent a home of their own. More often than not the first thing they do is apply a fresh coat of paint to freshen up the accommodation and give it their own identity.

Out of all the paint markets, the building trade is the biggest. The decorative market, together with the do-it-yourself market totals more than 60 percent of the paint volume in each country. All substrates can be coated with paint, such as: doors, window frames, ceilings, walls, etc. The remaining percentage is the protective coatings market: marine, automotive, car refinishing and industrial. All everyday objects that people use, are coated, such as: televisions, computers, shavers, lighting, etc.

There are many methods of paint application. Brushes, rollers and spray guns are the decorative and protective methods. For the industrial markets other applications have been developed such as: robot application, electrostatic application, dipping, etc. A pre-treatment is very important before each application. The different substrates demand different pre-treatments, such as wood, metal and concrete. After applying paint there is also the question – eventually – of how to remove it. Which removal methods are safe and environmentally friendly?

Adrie Winkelaar: Coatings Basics
© Copyright 2009 by Vincentz Network, Hannover, Germany
ISBN 978-3-86630-851-0

8.1　Application

The various paint applications in the various paint markets result from complex technology. The following paragraphs explain decorative, protective coating and industrial applications.

8.1.1　Decorative applications

Manual application of paint is the most usual method. Brushing is indispensable for small surface areas. In the decorative markets brushes and rollers are used to apply all kinds of paints, such as primers, lacquers, wall paints, etc.

The advantages of brushing are:

- low costs,
- minimum paint loss,
- better wetting and covering for any surface, and
- usable for any object.

The disadvantages or limitations of brushing are:

- labour-intensive,
- difficulty in achieving a uniform and precise film thickness and
- stripes from brush marks in critical-flow paints.

Figure 8.1: Painter with brush

Using rollers is much less labour-intensive than brushing and the film thickness is more uniform. Fine foam rollers with mohair provide the best results. For plaster and filler applications a filling knife is necessary.

The pre-treatment of any substrate is very important. Cleaning and sanding of old coatings is the first step. Untreated substrates need to have dust and other pollutions removed. Residues and loose, old coatings have to be removed.

Absorbing substrates must be pre-treated with primers and sealers.

The paint viscosity is of average value for an average application. Paint for large surfaces can be somewhat diluted with solvent or water. If the viscosity is too thin, the temperature could

Figure 8.2: Painter with roller

be too high. Placing the paint in a refrigerator should cool it down. For small brush strokes, such as for application to skirting boards or for the cut-in of window frames a higher viscosity is desirable. Cooling the paint down is a solution.

8.1.2 Protective coatings

The market for metal protection products is a huge professional market. Bridges, pylons, ships and steel constructions for buildings are very large surfaces for which protection and periodical maintenance are a necessity. Brushes, rollers and spray equipment are the usual methods of application. Compressed air guns are one of the most important application tools for this market. The advantages are:

- very good (fine) atomisation which results in a good surface quality,
- uniform thickness and an almost texture-free surface.

The disadvantages are:

- high material losses (overspray),
- high emission and
- spraying experience is essential.

High volume and low pressure (HVLP) equipment is a good solution to minimize the overspray. The principal of a spray gun is the spray chamber in which the paint supply and the air supply are combined. The outer nozzle provides the pneumatic atomisation. The diameters for the nozzle are:

- < 1.2 mm for low viscous materials
- 1.3 to 1.5 mm for high grade topcoats
- > 1.5 mm for intermediates, such as fillers and plasters

High pressure pneumatic equipment emits an atomising air of 2 to 7 bar and has a flow stream capacity of 0.2 to 0.8 m³/min. HVLP equipment has a pressure of 0.2 to 1.5 bar and a flow stream capacity of 0.5 to 2.5 m³/min. Hydraulic atomisation equipment provides very high pressure to the paint material, 80 to 400 bar and the air-assisted hydraulic atomisation (air-mix and air-less) provides a material pressure of 20 to 80 bar with an air pressure of 1.0 to 2.0 bar (see Table 8.1).

Table 8.1: Different spray methods

Equipment	Power of atomising air	Flow stream
High pressure pneumatic	2 to 7 bar	0.2 to 0.8 m³/min
HVLP	0.2 to 1.5 bar	0.5 to 2.5 m³/min
Hydraulic atomisation	80 to 400 bar	
Air-assisted hydraulic atomisation (air-mix and air-less)	20 to 80 bar (material pressure) 1.0 to 2.0 bar (air pressure)	

The supply of paint material is assured by the nozzle and is controlled with the injector by means of a trigger guard.

The pre-treatment of metals is very important. Cleaning and sanding the surface is the first step, with the activation of the surface against the oxidation as the second step by anti-corrosive primers. The third step is to protect the surface against external influences through the layer thickness. Additional layers are necessary to reach a thickness of 200 to 300 micron.

8.1.3 Industrial application

In the industrial paint market all possible techniques are applied. In addition to the above mentioned methods the following are also often applied:

- wiping,
- trowel application,
- curtaining,
- dipping,
- rotation atomisation and
- electrostatic powder coatings.

They distinguish between **wet** and **dry coating applications**. Wet coating applications are all viscous paints and dry coating applications are powder coatings. After any industrial paint application, fast drying equipment follows, such as stove and radiation curing.

Curtain coatings

Curtain coatings are used in the furniture industry to coat planed surfaces such as door panels and tabletops. The process is also applied in the metal and paper industries to coat metal and paper coils. The paint material is continuously circulated from the pouring tank to the collecting pan. Curtain coating is generally used in combination with radiation curing.

Dipping and flow coatings

These two simple coating methods are used for mass production articles and small parts of machinery. In conventional dipping processes in the joinery industry, window frames are immersed in the paint and dried on a rail system. Advantages of dipping are low application costs, minimum loss and very easily automated. In contrast to dipping and curtain coating, flow coating involves the pouring-over of the substrates contained in a closed compartment. The excess coating is collected and re-used.

Electrostatic atomisation

A special spraying method is the so-called electrostatic paint atomisation. The electrostatic field strength depends on the voltage and the distance between the work objects to be coated (positive charge) and the spray equipment (negative charge). In addition, the paint receives a negative charge. The advantages are: a uniform film thickness on the work objects – right up to the sharp tips and edges also.

The back spray effect also covers the back of the work objects and there is an excellent coating yield – almost 100% application efficiency. The disadvantage is the risk of excess coating at the edges.

Figure 8.3: Electrostatic spray equipment

Figure 8.4: Transit of metal strips through the powder coating cloud

Rapid-rotation atomisation

High rotation speeds from a rotation-disc (to 60,000 rpm). Very fine droplets result from the paint dropping down onto the rotating disc. Both this method and electrostatic atomisation make it possible for application to wood and plastic industrial work objects.

Powder coating application

Thermosetting powder (polyester, polyurethane and epoxy) and thermoplastic powder (acrylic) are applied to work objects using an electrostatic spraying process. Through stoving the powder fuses to form a sealed film and the thermosetting powder cures to become a hard sealed film. In the electrostatic powder spray gun the powder is fluidised with air. The powder particles are charged (negative) with the attachment of free electrons and air ions, which are generated in an applied high-voltage field together with one or more electrodes inside the gun. The high voltage generally reaches up to 90 kV to 120 kV. The advantages are: minimum emission, minimum coating loss, clean processing and very high coating quality.

Coil coating

The continuous coating of ships of steel or aluminium by mean of rollers is an important area of paint application. The use of coil coating metal sheet hat proven itself in the automotive industry and white goods industry.

8.2 Drying processes

During each drying process of a coating the volatile substances need to evaporate and the necessary chemical reaction must be take place. Chemical reactions occur faster at higher temperatures. There are four drying possibilities:

- drying at room temperature (air drying)
- forced drying up to approx. 80 °C (accelerated) in a cabin
- stove drying from 100 to 200 °C in a cabin
- radiation curing (UV light and IR light)

The **physical drying** of paints, e.g. cellulose lacquer, only dries through the evaporation of solvents. Solvents evaporate at all temperatures, even under 0 °C. The coalescent drying of paint is based on water, the evaporation of which is very slow. In wintertime it is possible to dry the wash outside, but under 0°C the wash and also the paint can freeze. The chemical reaction of the alkyd, epoxy and poly-urethane binders will cease when the temperature is less than 5 °C.

Forced drying that means a fast evaporation of the solvents or water through hot air circulation and the start of the chemical reaction of the binder if it is necessary. This solution takes place in a tunnel oven or in a cabin.

In **stove processing**, the paint on the treated objects is heated during a short time which enables film forming. The temperature depends on the binder type and varies between 100 and 200 °C. The air circulation during each drying process is very important. The circulating air flow must be 0.2 to 0.5 m/s.

Infra-red drying is very important for water based paints. Water evaporates very slowly and is dependant on the humidity in the air. A long-wave IR-lamp gives a temperature of 200 to 700 °C. The distance from the lamp and the drying time are very important for acquiring the correct film forming. Radiation curing is also depend-ant on the paint colour and brightness of the colour. Dark colours absorb more energy than light colours or metallic shades and dry faster with radiation curing.

UV light consists of electromagnetic waves with wavelengths rang-ing from 100 to 400 nm. This wavelength range is sometimes shorter than the violet visible light. This electromagnetic spectrum has a high-energy range and can be used for drying coatings. For the **UV drying** of paint, an UV initiator is necessary for the formation of free radicals, which results in a polymerisation reaction of the binder. After application, the UV systems are cured in a very short time, through irradiation, using UV light lamps.

8.3   Coating systems

All coating processes are expected to satisfy, as far as possible, the users' criteria of cost-effectiveness, quality – in regard to performance and appearance, and environmental compatibility. In each section of the market presents different requirements.

8.3.1 Decorative market

The decorative market is very subjective, because the developed norms and systems are unknown and each architect follows his or her own points of view and knowledge. A good pre-treatment, a good application of uniform film thickness and sufficient layers for protection are assets in different ways.

Professional painters undertake special education and are trained in how to apply different paints. On large surfaces e.g. doors, the painter applies the paint in thick brush strokes that result in the paint being well distributed to a surface of half a square meter. After applying a good uniform layer thickness, the painter applies once more to obtain a satisfactory end result. Paint is a part of a coating system: pre-treatment, equalizing the substrate with filler, afterwards a primer or sealer is necessary and finally the last layer is applied for decoration and protection.

8.3.2 Protective coating

The professional painter works according to international norms for the protection of steel substrates in various circumstances. The ISO 12944 states in C1 to C5 a qualitative class range for the application of the correct product for each corrosive circumstance. C1 and C2 are concerned with water-based products that are applied in light charged circumstances. C4 and C5 are concerned with solvent based products and are applied in heavily charged circumstances e.g. oilrigs, under water protection, etc.

8.3.3 Industrial coatings

In the automotive industry the optical requirements for surfaces is very important. Secondly, corrosion protection is an important part of the coating system. Steel sheets are treated with zinc phosphate and after modelling, the body is dipped into a cathode electro deposition coating. To reach a flat surface, surfacers and fillers are applied. Finally the applications of the topcoat or base coat-clear coat system are a part of the total coating system.

For electrical machinery special requirements have been developed for conductive or insulating. The coating system is built up in the same way. For optimal coating results a two- three- or four-coat system is recommended, depending on the technological and economical requirements.

8.4   Removal methods

In the paint industry there is extensive knowledge on the application of a coating systems, however, it is still not clear how to remove old coatings that have provided years of substrate protection. Knowledge regarding the old coating is essential. There is a difference between a physically dried product and a chemically dried product. The first type of coating can be removed using the correct solvent and the second type can only be treated through implementing a chemical coating removal method and a mechanical or thermal method.

8.4.1   Chemical coating removal

A chemical method is based on chemicals such as alkaline, acid or organic solvents. The most famous solvent for a paint stripper – methylene chloride – is forbidden in Europe. It is a very volatile and dangerous substance. In the trade markets, other products such as free of methylene chloride are available. The removal principle of these strippers is to cause the old coating to swell. The stripper must be applied on the old coating and after a certain time the coating becomes weak and can be removed with a scraper.

8.4.2   Mechanical coating removal

The old coating can be removed with abrasives or by grinding. The abrasives or grinding dust must be filtered under suction, because dangerous substances can be present in the old coating. Thick layers can be removed by coarse sandpapering and thin layers by fine sandpapering. Professional methods are blast cleaning and the use of a high-pressure water jet.

8.4.3   Thermal coating removal

In the trade market many hot-air blow dryers are available for the removal of old coatings. This method burns off the old coating, after which the residue can be removed with a scraper. The use of flame burners and open flames is forbidden because of the obvious danger. A new method is laser removal of old coatings.

Low temperatures are also used to remove old coatings. Embrittlement resulting from the use of liquid nitrogen at -196 °C is an industrial method of removing coatings from almost empty tins during the recycling process. Blasting with dry ice is also a professional thermal coating removal method.

9 Paint test methods

Paint test methods are very important for various reasons. In laboratories, during the development of paint, test methods are necessary to assess different proofs. During paint manufacturing test methods are usual for carrying out quality control. Test methods in the paint market are necessary for controlling the correct quality and correct application for the end user.

The paint testing people distinguish between **wet** and **dry** paints. For the wet paint

- soild content,
- viscosity,
- drying time,
- flow,
- film thickness,
- fineness

are important.

The solid content of wet paint is important for the filling properties and for protection. The viscosity and drying time is also important, such as the dust-free time and the press-free time.

For dried paint, the resistance against water, solvents and chemicals are important tests. Also

- hardness,
- flexibility,
- gloss,
- cleaning,
- hiding power,
- scratch resistance,
- abrasion resistance,
- durability and
- discolouring

are explained.

Adrie Winkelaar: Coatings Basics
© Copyright 2009 by Vincentz Network, Hannover, Germany
ISBN 978-3-86630-851-0

On the subject of paint as a very experimental material, everyone has an opinion. On this emotional subject the world test methods are more objective and provide the correct insight to the qualities of the various products. Without this insight it would be impossible to assess paints and coatings.

9.1   Wet properties

Wet properties are all of the properties of wet paint, both in tins and on substrates. This means that, the viscosity and drying time are the most important wet properties before the solvents evaporate and the binders react to become a hard and solid coating, In addition, knowledge should be acquired on the solid content, the specific gravity and fineness of the properties, and the paint quality.

Viscosity

Viscosity is a property of a liquid substance and expresses the resistance to forces against which substances can flow. If the ratio between the sheer rate and the sheer stress is constant, the result is an ideal liquid with a maximum flow. This liquid is called "**newtonian**" named after the English scientist *Newton*. Examples are: water, solvents and oil. If the ratio between the sheer rate and sheer stress is not constant the liquid is then called "**pseudo plastic**" or "**dilatants**".

The Newtonian liquids or paints can be measured with a flow-cup, according to ISO 2431. The content of the cup flows through a hole in the base of the cup within a specific time. Each cup is equipped with different nozzles (holes in mm). The larger the nozzle

Figure 9.1: Viscosity cup

the faster the liquid flows through. Very low viscous liquids are measured with nozzle number 2 or 4 and thick liquids are measured with nozzle number 6.

Many paints are pseudo plastic, such as thixotropic, and are measured with rotational viscometers. A rotary element swirls in a sample container and measures the resistance (sheer stress) in the liquid. A thixotropic paint appears to be thick, but after stirring the paint appears to be much thinner.

Figure 9.2: Rotational viscose meter

Drying time

Tests on drying properties are carried out on a wet paint film over a period of time. A needle connected to a clock moves through the wet paint film making specific drying patterns, for example, stripes that

Figure 9.3: Drying recorder for wet paint on glass strips

flow together. After a given period of time the stripes cease to flow together (this is the first step in the drying process) and after another period of time the needle makes a mark in the sticky paint film (this is the second step). When the coating is completely dry the needle no longer affects the paint film (this is the final step in the drying process). The dryness test is carried out on glass strips maintaining a constant layer thickness, under a constant temperature and in relatively humid conditions.

Flow and sagging

Flow is defined as the capacity of a still liquid film compensating independently any irregularities that have been introduced during application, such as brush stripes or spray effects. Running, sagging and dripping are negative properties of the same flowing process. The slower the drying process (chemical and physical drying) the more time is available for the flowing process. A test blade for measuring the flow and running behaviour contains five pairs of gaps of 1.0 mm to 0.1 mm. With this test blade a wet paint film is applied on glass strips, which are placed vertically to dry. After several minutes it is possible to see which layer thicknesses are sensitive to sagging.

Solid content

The solid content of paint provides an indication for filling and protection properties. The non-volatile matter is the residue obtained through evaporation under fixed conditions (ISO 4618). All pigments, extenders, matting agents and binders are solid particles in paint. The solvents, including water, are the volatile components. A small percentage of additives are partly solid and partly volatile. The solid content is a percentage of the total wet paint.

Specific gravity

The specific gravity is expressed in grams per litre (volume) of a product. This means that the ratio between volume and weight provides an indication of heavy substances in paint, such as pigments and extenders. Chapter 2.3 and Chapter 5.1 mentioned that all pigments and extenders have different densities: between 1.5 and 5.0 g/cm^3. All binders have a density of 1.2 g/cm^3. Water is 1.0 g/cm^3 and most solvents vary between 0.7 and 1.0 g/cm^3. When a product has a specific gravity of 1.3 g/cm^3 and the spreading rate is 10 m^2 per litre, the weight is 1.300 g/m^2. The wet layer thickness would be, in that case 0.1 mm (100 micron)

Wet film thickness

A simple method for measuring the wet film thickness is with the use of a comb that has teeth of varying lengths for example 50, 100, 150, 200 and 250 micron. When the comb is pressed into the wet paint layer and only the teeth of 50 and 100 micron have touched the paint leaving the 150, 200 and 250 micron teeth clean, then the wet layer is, in that case, between 100 and 150 micron. The comb (called the Rossmann comb) can have up to three sides with different scales of layer thickness.

There is also modern radiometric film thickness measuring techniques including X-ray fluorescence methods or beta backscattering methods. These are applied in industrial procedures for, amongst other things, coil coatings and automotive assembly lines.

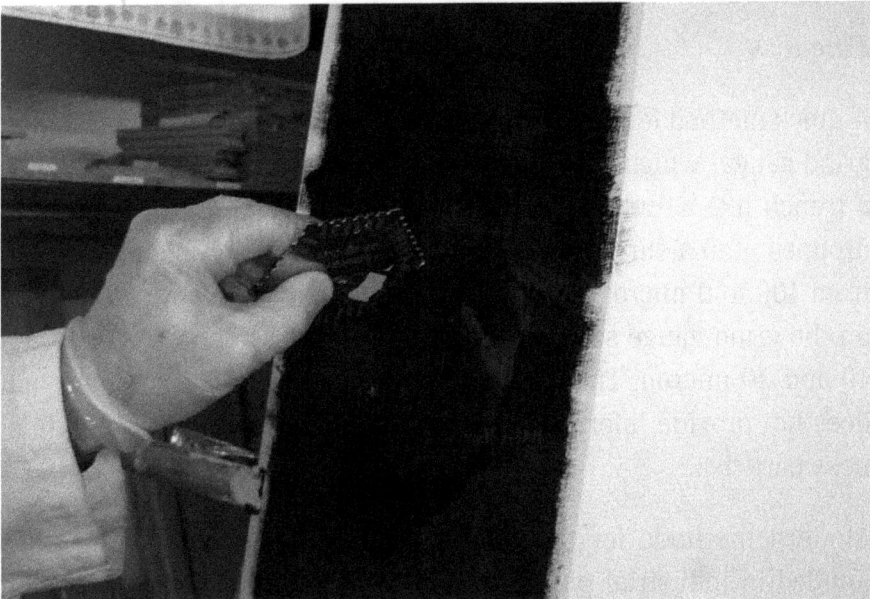

Figure 9.4: Measuring wet film thickness with comb

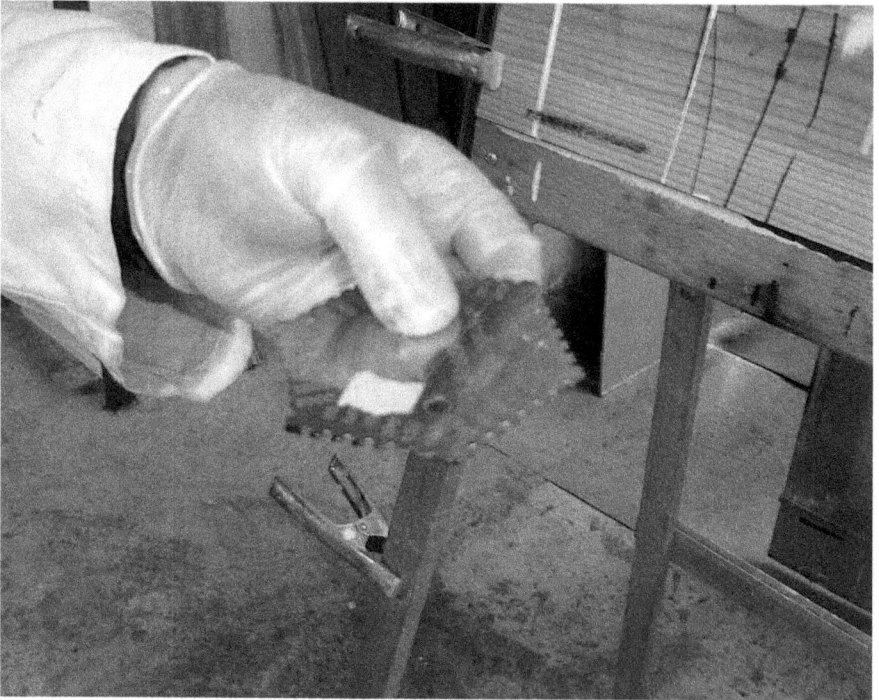

Figure 9.5: Reading the wet film thickness

Fineness

A quick method for measuring fineness is with the use of a Hegman grind-gauge, which is described in ISO 1524. The grind gauge has a trench like a "wedge" of 100 to 0 micron. When a little paint is dropped onto a surface of 100 micron and a 'doctor' blade is pulled from 100 to 0 micron, the correct fineness of the paint can be read on the grind gauge scale. Normally paint has a fineness of between 10 and 20 micron. This gauge only measures the top fineness and does not provide information regarding the distribution of the fineness particles.

Modern methods for determining particle size distribution are applied in industrial processes with the use of photon correlation spectroscopy and laser light diffraction.

Figure 9.6: Hegman grind-gauge without paint

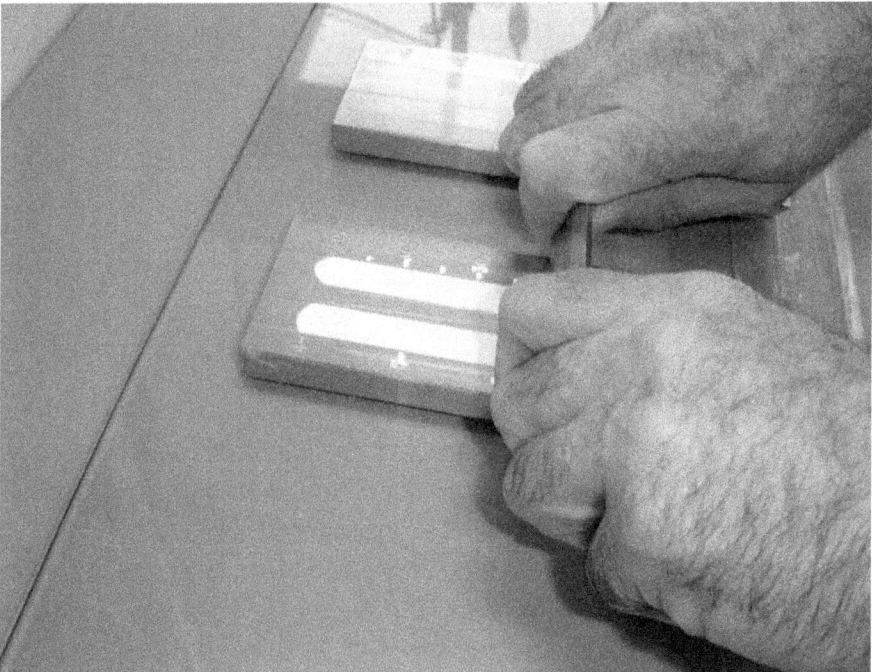

Figure 9.7: Applying paint on the Hegman grind-gauge

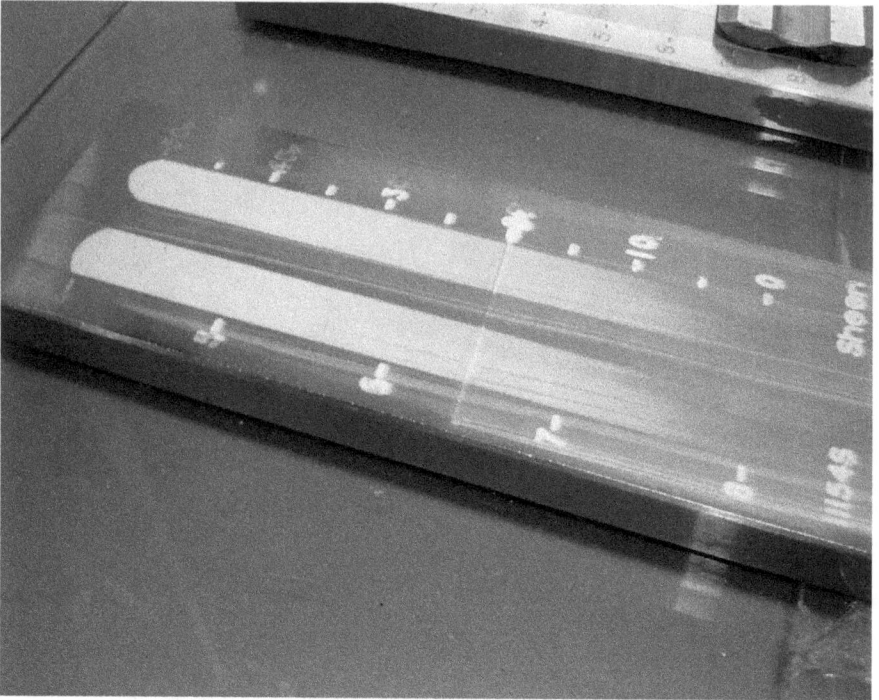

Figure 9.8: Reading the fineness on the Hegman grind-0-gauge

9.2   Dried properties

Most measurement tests are carried out on the dried paint film: "the coating". Primary properties are

- gloss,
- adhesion,
- hardness,
- resistance and
- durability.

However, each application has its own specific properties. Coatings for doors must be scratch resistant and must be able to be cleaned at any time. Coatings for walls must have the same properties, but not in the same range. Exterior window frame coatings must also have the same properties in addition to coatings for toys, garden machinery, computers and oil refineries. The amount of resistance and the resistance to cleaning fluids that are used in the application are just as important. In addition durability varies enormously for numerous

Figure 9.9: Gloss reflectometer

paint applications. There are enormous differences between interior and exterior applications and between decorative and industrial applications also.

Gloss

Gloss is measured using reflectometers ISO 2813 by which the angles of incidences can be selected according to the gloss character of the coating film. Directed incident light is reflected by the spread properties of the film. For moderate degrees of gloss (30 to 70 gloss units) the measuring angle is 60°. For high gloss surfaces (more than 70 gloss units at 60° geometry) a 20° measuring angle is recommended. For matt surfaces (under the 30 gloss units at 60° geometry) an 85° measuring angle is advised.

Film thickness

The methods for measuring the thickness of dry coatings can be divided into "destructive" and "non-destructive". This means that destructive methods cause damage to coatings. The principle is to make a V-shaped

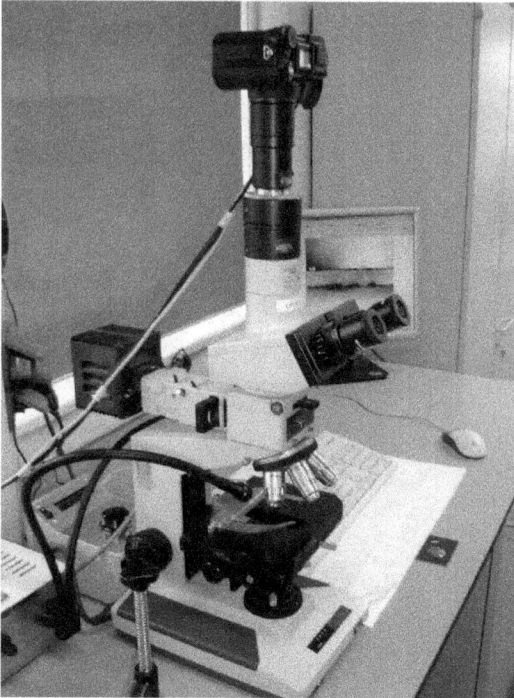

Figure 9.10: Microscope for measurement film thickness

cut or other damage in the coating and measure the thickness optically or mechanically. The sample must be a perpendicular incision to the surface that can be measured under a light microscope.

A non-destructive method is based on the magnetic adhesion when the coating is applied to ferro metals. Depending on the thickness of the intermediate coating film the adhesion to iron provides the thickness on a scale.

Adhesion

The cross-hatch adhesion test ISO 2409 is performed with a multi-blade knife comprising 10 sharp knifes with distances of 1 to 2 mm. With this multi-blade knife a cut in the coating film can be made in a random direction. After that a second cut must be made at the right angle to the first. The adhesion measures are then tested using sticky tape which is pressed onto the hundred block cuts in the coating film and then pulled away. In practice, a knife or sticky tape is often used to scratch or pull a coating film from a substrate.

The adhesive strength can also be measured with the 'pull-off' test using a disc bonded to the surface and a tensile testing machine ISO 4624. This test must be performed with great care.

Hardness, scratch resistance and abrasion

Hardness is one of the most difficult properties to describe. Softness can be caused by too short drying time or by too much power damaging the coating. The deformability or elasticity has the tendency to reverse a deformation caused by an external influence. Hardness can be defined as the resistance of the coating film to outside forces. This indentation hardness can be measured using the 'Buchholz' method and is described in ISO 2815. A conical point is pressed into a coating film with an indenter during a standard given time. The measurement provides the hardness according to Buchholz.

The **scratch hardness** is tested with a cutter (a stylus) applying increasing indentation force until a permanent line appears on the surface (ISO 1518). The weight is the measure of the scratch resistance. The pencil test using various pencils of different degrees of hardness (6B-B-HB-2H-9H) can also be applied to measure the scratch hardness.

Figure 9.11: Buchholz measurement (Erickson)

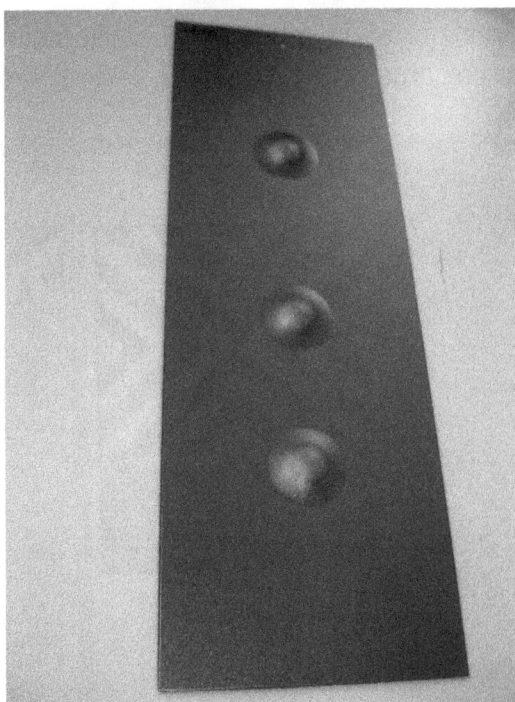

Figure 9.12: Test panel with Buchholz measurements

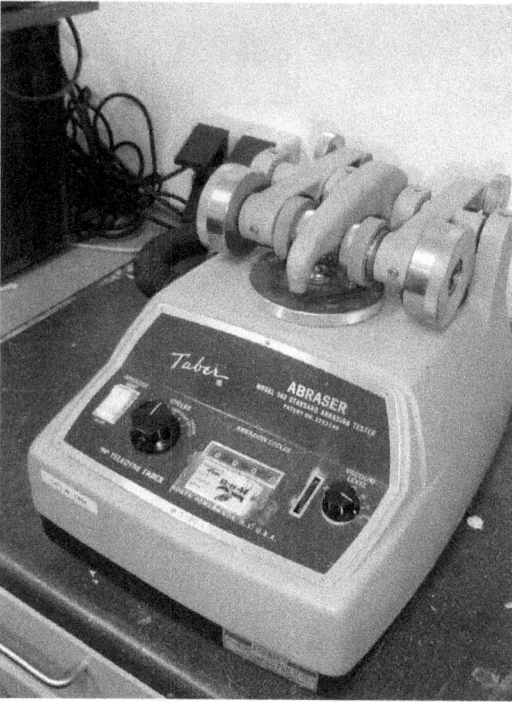

Figure 9.13: Taber Abraser (abrasion test)

The **abrasion resistance** of a coating is generally tested by means of surface abrasion using sandpaper or sand. The Taber Abraser comprises two abrasing wheels which rotate on the turning test pieces. This apparatus is used to test parquet varnishes.

Hiding power

The hiding power is the capacity of a coating for concealing a substrate and is determined by the comparative measurements of the reflection of a coating on a black and white substrate or on cards (ISO 2814, 3905, 3906 and 6504).

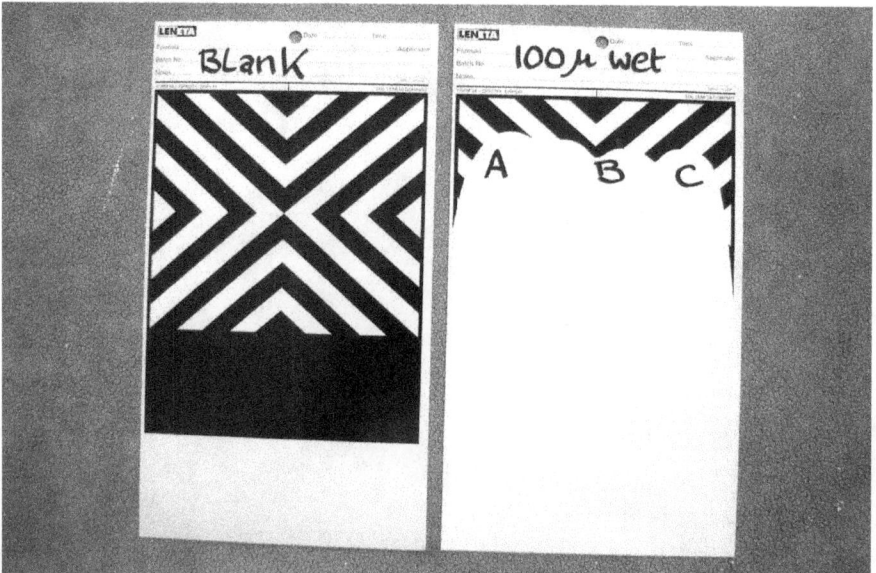

Figure 9.14: Hiding power: test on contrast cards

The hiding power value indicates the surface area (m²) of a contrasting substrate that can be covered with a litre of dry coating. The brightness can be measured using a photometer on the black surface and on the white surface. Both values provide the hiding property values. As higher the value as better is the hiding power.

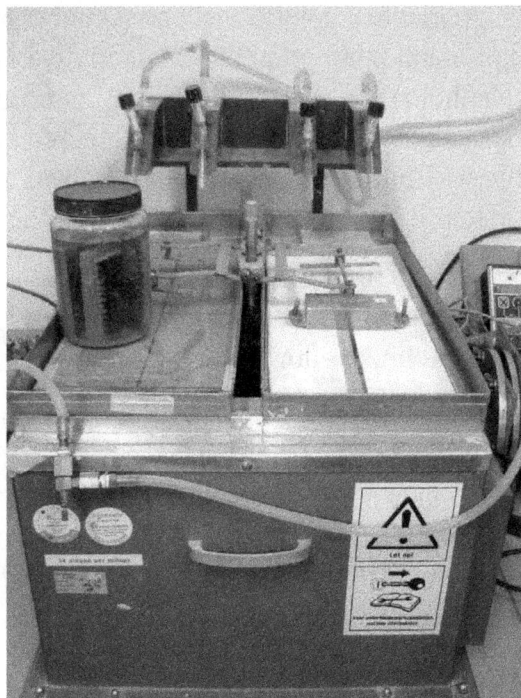

Figure 9.15: Cleaning test

Cleaning resistance

Cleaning refers to the removal of any foreign bodies/substances from a coating surface using cleaners and solvents. Not all coatings have the same resistance to cleaners. This depends on the binder type and the binder amount in the paint formula. A test method for cleaning is to rub the coating surface with a cloth saturated with cleaner. If the coating dissolves and starts to run on the surface, then the cleaner is too aggressive. More water and soap are necessary to clean the surface without damaging it.

The Gardner scrubbing tester is used to determine washing and scrubbing resistance to a coated surface. Together with a standard scrubbing/cleaning fluid, two brushes move backwards and forwards on test panels. After a defined number of scrubbing cycles, the damaged pattern is evaluated. The number of cycles decides the quality of the resistance to cleaning.

Colour and discolouring

Chapter 4.3 mentions that all colours can be measured by photo-electric cells. How do we make the leap from light reflection of a coloured test panel to a colorimetric system of coordinates that correspond to our visual perception? The sensitivity of the human eye depends on the photo pigments of the retinal cones in the eyes. The light intensities, allocated to three ranges: blue, red and green, are also known as the tristimulus values. The best known and most widely used colour system is the CIE-Lab system, in which

- L is the brightness (ideal white);
- a is green-red component and
- b the blue-yellow component.

Colour differences between a reference and a sample can be expressed in ΔL, Δa and Δb (Δ is the sign in mathematics for the differences). The same applies for colour saturation and the hue of a colour. The overall colour difference can be expressed in ΔE. The colour tolerance to assess colour differences lies in the range of ΔE = 0.5 to 1.0.

Corrosion test

Corrosion protection tests are for the prevention/protection of corrosion to coatings on metal substrates. Many anti-corrosion coatings exhibit a retarded effect through anti-corrosion pigments such as zinc, phosphate, etc. In corrosive surroundings dampness and the presence of salt have much influence on the corrosive process. The "salt spray test" (ISO 7253) is the best known test for anti-corrosive coatings on metal panels. It consists of a continuous spraying of a salt solution within a salt spray chamber and at an elevated temperature for a period of 3 to 10 days. The surface of a

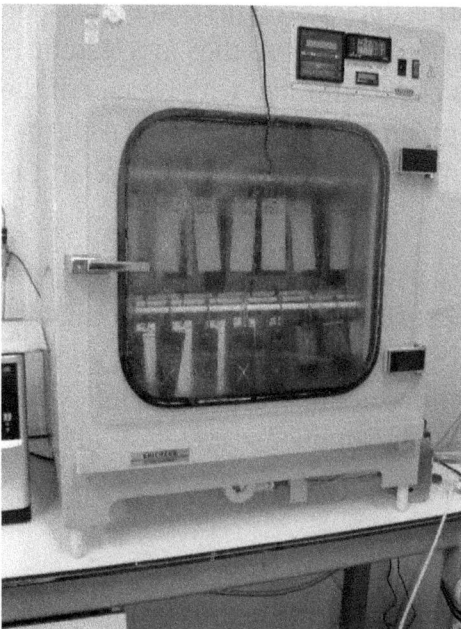

Figure 9.16: Corrosion test

test panel is usually indented by cross-cutting in the coating surface through to the metal substrate. The test is carried out in accordance with ISO 4628 for the evaluation of rust formation and blistering.

Weather resistance

In order to test coatings against the weather influences, test pieces must be prepared for, and exposed to extreme weather conditions. Many influences such as UV radiation, salt-laden sea air, humid and warm tropical climates with enormous temperature differences between day and night and influences of an industrial atmosphere, can cause damages to the coating protection. Normally, test pieces are exposed in a field-rack so that comparisons can be made under the same conditions. The angle of 45° in direction of south-west is usually implemented for testing coating surfaces against UV radiation and discolouring. The 90° angle is used for humidity and rain influence to window frames, because the horizontal parts show the first signs of damage to the coating system.

Figure 9.17: Weather resistance equipment (QUV)

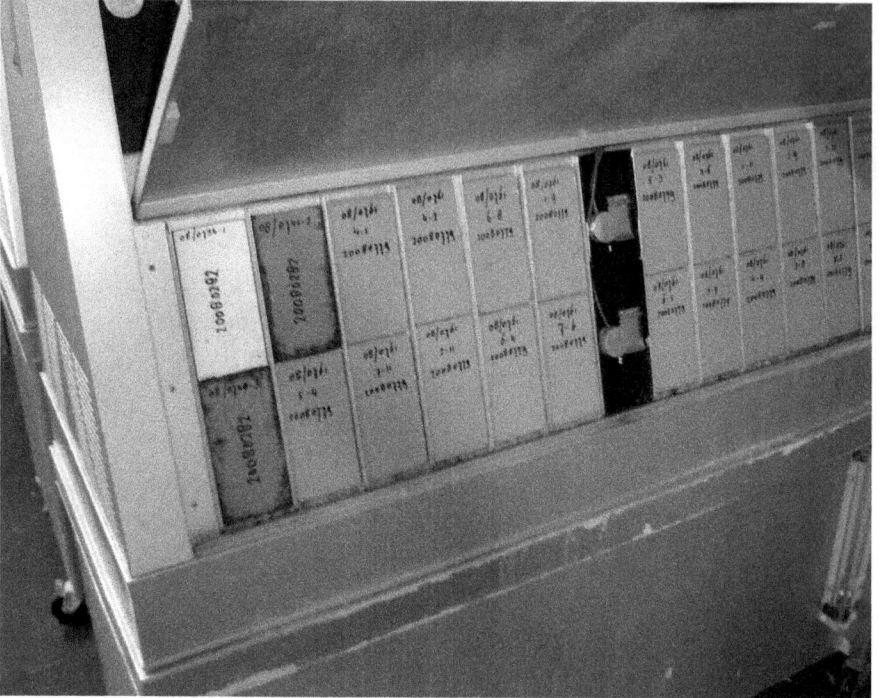

Figure 9.18: Weather resistance (open with test panels)

To avoid having to expose products for many years, accelerated results are required. The reduction factor for solar radiation in Florida is around 50%, which means that one year's exposure in Florida is the equivalent of approximately two years in central Europe.

One of the best pieces of equipment for accelerated weathering is the "weather-o-meter" (WOM). The pieces to be tested are placed in the inside drum of the WOM that passes UV-lamps and sunshine-lamps while rotating. In a cycle of day-night and humid-dry an indication can be obtained regarding durability.

A weather test is often followed by a chalk test together with a gloss retention test. The chalking and gloss tests are carried out by using sticky tape that is applied to the exposed coating on a black substrate to show the degree of chalking

10 Health, safety and environment

Health, safety and environment (HSE) regulations regarding paint are becoming increasingly important, because paint consists of reactive substances that enable drying, adhesion, hardness build-up and durability. On principle each type of paint, including natural based paint consists of dangerous properties, therefore risks to health and the environment are always present.

Drying properties, adhesion and block properties can damage living organisms. A 'Material Safety Data Sheet' (MSDS) for chemicals and products containing chemicals is worldwide obligatory for professional painters. In the DIY market the sales people are obliged to inform the paint user about the product risk. In each European MSDS there are 16 categories that provide information concerning dangerous substances, safe handling, exposure time, emissions, pollution control of air, water and soil, safe transport and storage. European legislation is the starting point of this chapter.

10.1 Pollution control (air, water, soil and packages)

Air pollution, water pollution and soil pollution must be avoided. The legislation defines each kind of pollution caused by dangerous chemical substances. There is a difference between wet paint and dried coatings. All wet paints are dangerous substances and must be separately collected and disposed as chemical waste. Dried coatings can be disposed as household waste, however, the dried coating is considered dangerous when there are for example heavy metals present and must therefore be considered a chemical waste.

Adrie Winkelaar: Coatings Basics
© Copyright 2009 by Vincentz Network, Hannover, Germany
ISBN 978-3-86630-851-0

With regard to air pollution, the European Emission Directive has been adapted to, among others, VOC (Volatile Organic Compounds). VOC substances are present in many paint types and for industrial application there are regulations necessary in order to restrict the emission using filters. For decorative application a Product Directive is adapted (2004/42/EU) to reduce the amount of VOC in the paint products.

Water pollution legislation is a part of the Global Harmonised System of Classification and labelling of Chemicals and Dangerous Goods. The classification for contents that are dangerous to the environment is the use of a symbol depicting "a dead tree and a dead fish". Paint chemicals must never be deposited in a sewerage system. This means that before the waste water can be disposed of, a water cleaning machine is necessary.

The same explanation is in force for soil pollution. Paint contents can cause soil pollution because paint is a hardening substance that may also contain heavy metals. Each year, the paint industry worldwide,

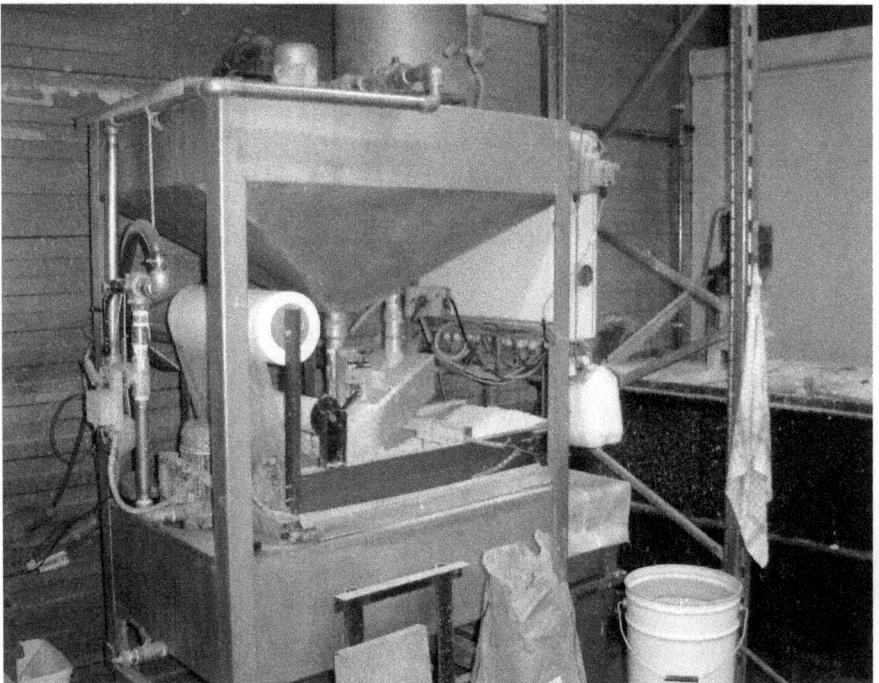

Figure 10.1: Water cleaning machine with flocculate agent and filtration cloth

produces millions of tins and buckets of paint in each country. In nearly every kitchen cupboard, attic, shed or garage – tins containing left over paint can be found. Paint packages are produced very accurately because paint is a dangerous substance. The package must be safe during transport and storage. Empty tins without wet paint are considered household or company waste if no heavy metals or biocides remain in the paint. Left over paint packages are dangerous or chemical waste.

A good collecting system in each country of empty packages and chemical waste is necessary. Therefore paint packages can be recycled. Old tins are remade into new tins and plastic buckets are used to supply new energy in a waste utilization plant.

10.2 Safe handling according to Material Safety Data Sheet

Paint materials can cause two types of dangers:

- First: a risk of fire and explosions and
- Second: a risk to health and environment.

Each Material Safety Data Sheet (MSDS) must describe all of the properties (see Table 10.1, page 126).

The hazards identification contains information about skin contact, skin absorption and eye contact. Inhalation information and ingestion information is also necessary. The system of classification and labelling of chemicals and dangerous goods has been devised worldwide.

The following labelling symbols (page 126) are used in the European Union, the European Economic Area and some other countries. They were originally defined in Annex 2 of the Directive 67/548/EEC and have been recently published in the Official Journal for the European Communities L 225, 21/08/2001 pp. 1–333

Each paint product has to carry a label with symbols and Risk and Safety phrases of the same European Directive (see Appendix 3). Additionally, the directions for use or Technical Data Sheets and HSE-information can be found on each paint package.

Table 10.1: Sections of the European Material Safety Data Sheet (MSDS)

Section 1	Chemical product and company identification	Section 9	Physical and chemical properties
Section 2	Information of hazardous ingredients	Section 10	Stability and reactivity
Section 3	Hazard identification including emergency overview	Section 11	Toxicological information
Section 4	First aid measures	Section 12	Ecological information
Section 5	Fire fighting measures	Section 13	Disposal consideration
Section 6	Accidental release measures	Section 14	Transport information
Section 7	Handling and sto-rage	Section 15	Regulatory information
Section 8	Exposure controls/personal protection	Section 16	Other information

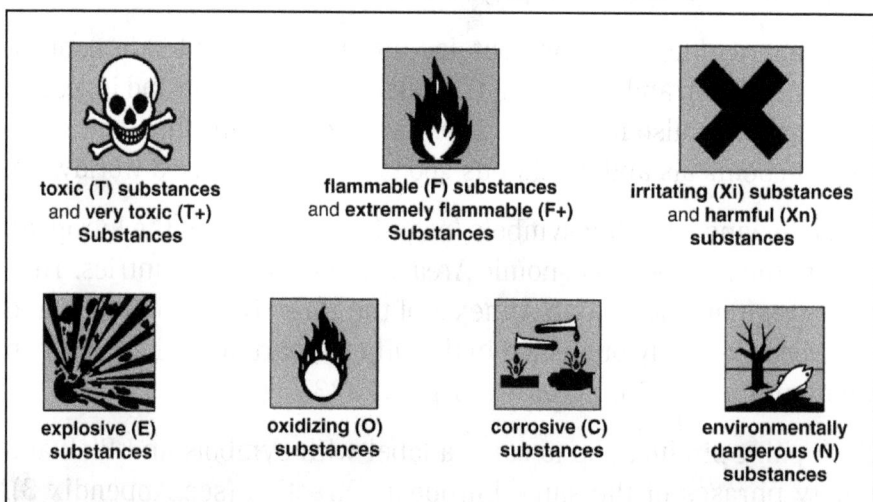

toxic (T) substances and very toxic (T+) Substances

flammable (F) substances and extremely flammable (F+) Substances

irritating (Xi) substances and harmful (Xn) substances

explosive (E) substances

oxidizing (O) substances

corrosive (C) substances

environmentally dangerous (N) substances

Figure 10.2: Labelling symbols as a system of classification and labelling of chemicals and dangerous goods

10.3 Exposure and emissions

In section 8 of the MSDS, information can be found on practice or equipment or on both. This is extremely useful for the minimization of exposure of the user. In this section the correct personal protective equipment also has to be stated. Toxicological and ecological information has to be stated in sections 11 and 12. This provides information regarding the effect of acute exposure and chronic exposure.

To reduce the VOC (volatile organic compounds) amount in decorative and refinishing paint products the Product Directive 2004/42/E states the limits of 12 decorative product groups. In 2010 the following limits will be obligatory in Europe. See Table 10.2.

Table 10.2: VOC limits of decorative products

Product subcategory	Waterbased in g/l VOC	Solventbased in g/l VOC
a) Interior matt walls and ceiling (Gloss < 25 GU at 60°)	30	30
b) Interior glossy walls and ceilings (Gloss > 25 GU at 60°)	100	100
c) Exterior walls of mineral substrate	40	430
d) Interior/exterior trim and cladding paints for wood and metal	130	300
e) Interior/exterior trim varnishes and wood stains, including opaque wood stains	130	400
f) Interior and exterior mineral built wood stains	130	700
g) Primers	30	350
h) Binding primers	30	750
i) One-pack performance coatings	140	500
j) Two-pack reactive performance coatings for specific end use such as floors	140	500
k) Multi-coloured coatings	100	100
l) Decorative effect coatings	200	200

For vehicle refinishing products, the limits that were issued in 2007 apply to 5 product groups. See Table 10.3.

Table 10.3: Maximum VOC content limit value for vehicle refinishing producers

Product subcategory	Coatings	g/l VOC
Preparatory and cleaning	preparatory pre-cleaner	850 200
Body filler/stopper	all types	250
Primer	surfacer/filler and general (metal) primer wash-primer	540 780
Topcoats	all types	420
Special finishes	all types	840

Since 2000 only waterbased paint products are obligatory in The Netherlands in order to reduce exposure risk to professional painters working with interior applications. This obligation has been extorted by the Trade Unions under local legislation concerning workloads.

10.4   Transport and storage

All regulations for the transportation of dangerous goods has been harmonised globally. These regulations comprise 9 classes. Flammable paint has been divided up in class 3 and some miscellaneous products in class 9 (see Table 10.4). Many paint products are exempt because the risks of calamities are minimal.

Road transport (ADR), railway transport (SRA), air transport (IATA) and water transport (IMO) including maritime and shipping apply the same principles, based on the regulations of dangerous substances. There are numerous rules for packaging, classes of dangerous goods and transport security that apply to the transportation of paint.

Table 10.4: Transportation classes: regulations for the transportation of dangerous goods

Class 1	explosives
Class 2	gasses
Class 3	flammable liquids
Class 4	flammable solids
Class 5	oxidizing substances and organic peroxides
Class 6	toxic and infectious substances
Class 7	radio active materials
Class 8	corrosives
Class 9	miscellaneous products/substances or organisms

Storage regulations have been adapted for toxic, irritating, flammable and environmentally dangerous paint products. Fire-fighting water has to be collected and should not flow into the sewerage systems or directly into the environment.

References

Literature

Dr. Th. Brock, Dr. M. Grotekleas and Dr. P. Mischke, European Coatings Handbook, Vincentz Network 2009

Dr. B. Müller and U. Poth, Coatings Formulation, Vincentz, Network 2006

Dr. Ulrich Meier-Westhues, Polyurethanes for Coatings: Adesives and Sealants, Vincentz Network 2007

Dr. R. Baumstark and Dr. M. Schwartz, Waterborne Acrylates, Vincentz Network 2001

P. G. de Lange, Powder Coatings, Chemistry and Technology, Vincentz Network 2004

D. Satas and A. D. Tracton, Coatings Technology Handbook, Marcel Dekker 2001

Internet

www.coatings.de

www.cepe.org

www.chemnet.com

www.european-coatings.com

www.pra.uk

Adrie Winkelaar: Coatings Basics
© Copyright 2009 by Vincentz Network, Hannover, Germany
ISBN 978-3-86630-851-0

Appendix 1

Periodic table of elements

Atomic number	Element symbol	Element	Atomic number	Element symbol	Element
1	H	Hydrogen	18	Ar	Argon
2	He	Helium	19	K	Potassium
3	Li	Lithium	20	Ca	Calcium
4	Be	Beryllium	21	Sc	Scandium
5	B	Boron	22	Ti	Titanium
6	C	Carbon	23	V	Vanadium
7	N	Nitrogen	24	Cr	Chromium
8	O	Oxygen	25	Mn	Manganese
9	F	Fluorine	26	Fe	Iron
10	Ne	Neon	27	Co	Cobalt
11	Na	Potassium	28	Ni	Nickel
12	Mg	Magnesium	29	Cu	Copper
13	Al	Aluminium	30	Zn	Zinc
14	Si	Silicon	31	Ga	Gallium
15	P	Phosphorus	32	Ge	Germanium
16	S	Sulphur	33	As	Arsenic
17	Cl	Chlorine			

Adrie Winkelaar: Coatings Basics
© Copyright 2009 by Vincentz Network, Hannover, Germany
ISBN 978-3-86630-851-0

Atomic number	Element symbol	Element	Atomic number	Element symbol	Element
34	Se	Selenium	51	Sb	Antimony
35	Br	Bromine	52	Te	Tellurium
36	Kr	Krypton	53	I	Iodine
37	Rb	Rubidium	54	Xe	Xenon
38	Sr	Strontium	55	Cs	Cesium
39	Y	Yttrium	56	Ba	Barium
40	Zr	Zirconium	57	La	Lanthanum
41	Nb	Niobium	58	Ce	Cerium
42	Mo	Molybdenum	59	Pr	Praseodymium
43	Tc	Technetium	60	Nd	Neodymium
44	Ru	Ruthenium	61	Pm	Promethium
45	Rh	Rhodium	62	Sm	Samarium
46	Pd	Palladium	63	Eu	Europium
47	Ag	Silver	64	Gd	Gadolinium
48	Cd	Cadmium	65	Tb	Terbium
49	In	Indium	66	Dy	Dysprosium
50	Sn	Tin			

Atomic number	Element symbol	Element	Atomic number	Element symbol	Element
67	Ho	Holmium	86	Rn	Radon
68	Er	Erbium	87	Fr	Francium
69	Tm	Thulium	88	Ra	Radium
70	Yb	Ytterbium	89	Ac	Actinium
71	Lu	Lutetium	90	Th	Thorium
72	Hf	Hafnium	91	Pa	Protactinium
73	Ta	Tantalum	92	U	Uranium
74	W	Tungsten	93	Np	Neptunium
75	Re	Rhenium	94	Pu	Plutonium
76	Os	Osmium	95	Am	Americium
77	Ir	Iridium	96	Cm	Curium
78	Pt	Platinum	97	Bk	Berkelium
79	Au	Gold	98	Cf	Californium
80	Hg	Mercury	99	Es	Einsteinium
81	Tl	Thallium	100	Fm	Fermium
82	Pb	Lead	101	Md	Mendelevium
83	Bi	Bismuth	102	No	Nobelium
84	Po	Polonium	103	Lr	Lawrencium
85	At	Astatine			

Appendix 2

Old fashion ingredients and formulations

In the past, paint makers used every raw material that they could find to create a good paint application. **Lead**, **cadmium** and **chromate pigments** were used abundantly because the hiding power was excellent and at that time the raw material prices were low. Solvents were also used to make paint cheaper. In combination with thickeners, many solvents could be added so that the litre or kilo price could be reduced. At present toxic and/or harmful substances are ever decreasing as a result of the strict legislations (REACH) in Europe which make it difficult to continue to use certain substances. REACH means Registration, Evaluation, Authorization and Restriction of Chemicals and was introduced in 2008. In future years (up to 2018) registration of all chemicals will be compulsory.

As background information, you will find some explanatory information regarding the ingredients and formulations of 'old fashioned' substances listed below.

Pigments

Cadmium pigments

Cadmium (Cd = 48) pigments were manufactured with sulphide (S = 16) and selenide (Se = 34). They were characterised by particularly brilliant shades in yellow, orange and red. Also the temperature stability and migration resistance was very good. Nowadays because cadmium is very toxic the pigments can only be used if there is no adequate alternative substitution possible.

Adrie Winkelaar: Coatings Basics
© Copyright 2009 by Vincentz Network, Hannover, Germany
ISBN 978-3-86630-851-0

Chromate and lead pigments

Pigments in this group are also known as lead (Pb = 82) chromates and exist in the colour ranges of yellow, orange and red. Chromium (Cr = 24) yellow, orange, red and green (in combination with iron blue) and molybdate (Mo = 42) the red and orange colours are very hazardous to health and environment. They have carcinogenic potential and are toxic in reproduction. Lead compounds and preparations thereof with very little lead content require special labelling.

Formulations

Nitrocellulose

When cellulose, a component of soft wood, reacts with acid (nitric and sulphuric acid), nitrocellulose is formed which is soluble in alcohol and other solvents. Plasticisers are necessary to enable more flexible films. Cellulose lacquers have 30 to 40 % solid content which means more than 60 % solvents.

Chlorinated rubber

Natural rubber is produced from the rubber tree. The milky sap, called latex, was used in the first latex wall paints in the middle of the last century. Modern synthetic polymers are increasingly replacing this natural substance. Chlorine rubber is caused by reactions from chlorinated substances and is applied as a corrosion protective coating. These physical drying products also contain a lot of solvents and are being used less and less.

Appendix 3

Labbeling phrases

Chemical data sheets available in many countries now contain codes for certain "risk phrases", shown as R23, R45 etc. These risk-phrase codes have the following meanings:

R1	Explosive when dry
R2	Risk of explosion by shock, friction, fire or other source of ignition
R3	Extreme risk of explosion by shock, friction, fire or other sources of ignition
R4	Forms very sensitive explosive metallic compounds
R5	Heating may cause an explosion
R6	Explosive with or without contact with air
R7	May cause fire
R8	Contact with combustible material may cause fire
R9	Explosive when mixed with combustible material
R10	Flammable
R11	Highly flammable
R12	Extremely flammable
R13	Extremely flammable liquefied gas
R14	Reacts violently with water
R15	Contact with water liberates extremely flammable gasses
R16	Explosive when mixed with oxidizing substances
R17	Spontaneously flammable in air
R18	In use, may form inflammable/explosive vapour-air mixture
R19	May form explosive peroxides
R20	Harmful by inhalation
R21	Harmful in contact with skin

Adrie Winkelaar: Coatings Basics
© Copyright 2009 by Vincentz Network, Hannover, Germany
ISBN 978-3-86630-851-0

R22	Harmful if swallowed
R23	Toxic by inhalation
R24	Toxic in contact with skin
R25	Toxic if swallowed
R26	Very toxic by inhalation
R27	Very toxic in contact with skin
R28	Very toxic if swallowed
R29	Contact with water liberates toxic gas
R30	Can become highly flammable in use
R31	Contact with acids liberates toxic gas
R32	Contact with acid liberates very toxic gas
R33	Danger of cumulative effects
R34	Causes burns
R35	Causes severe burns
R36	Irritating to eyes
R37	Irritating to respiratory system
R38	Irritating to skin
R39	Danger of very serious irreversible effects
R40	Limited evidence of a carcinogenic effect
R41	Risk of serious damage to the eyes
R42	May cause sensitization by inhalation
R43	May cause sensitization by skin contact
R44	Risk of explosion if heated in confinement
R45	May cause cancer
R46	May cause heritable genetic damage
R47	May cause birth defects
R48	Danger of serious damage to health by prolonged exposure
R49	May cause cancer by inhalation
R50	Very toxic to aquatic organisms
R51	Toxic to aquatic organisms
R52	Harmful to aquatic organisms
R53	May cause long-term adverse effects in the aquatic environment
R54	Toxic to flora
R55	Toxic to fauna
R56	Toxic to soil organisms

R57	Toxic to bees
R58	May cause long-term adverse effects in the environment
R59	Dangerous to the ozone layer
R60	May impair fertility
R61	May cause harm to the unborn child
R62	Risk of impaired fertility
R63	Possible risk of harm to the unborn child
R64	May cause harm to breastfed babies
R65	Harmful: may cause lung damage if swallowed
R66	Repeated exposure may cause skin dryness or cracking
R67	Vapours may cause drowsiness and dizziness
R68	Possible risk of irreversible effects
R20/21	Harmful by inhalation and in contact with skin
R20/21/22	Harmful by inhalation, in contact with skin and if swallowed
R20/22	Harmful by inhalation and if swallowed
R21/22	Harmful in contact with skin and if swallowed
R23/24/25	Toxic by inhalation, in contact with skin and if swallowed
R23/25	Toxic by inhalation and if swallowed
R26/27/28	Very toxic by inhalation, in contact with skin and if swallowed
R26/28	Very toxic by inhalation and if swallowed
R36/37	Irritating to eyes and respiratory system
R36/37/38	Irritating to eyes, respiratory system and skin
R36/38	Irritating to eyes and skin
R37/38	Irritating to respiratory system and skin
R42/43	May cause sensitization by inhalation and skin contact
R48/22	Harmful: danger of serious damage to health by prolonged exposure if swallowed
R50/53	Very toxic to aquatic organisms, may cause long-term adverse effects in the aquatic environment
R51/53	Toxic to aquatic organisms, may cause long-term adverse effects in the aquatic environment
R52/53	Harmful to aquatic organisms, may cause long-term adverse effects in the aquatic environment

Author

Adrie Winkelaar was born in 1946. After his study Chemical Technology he worked 25 years in the AkzoNobel research department Decorative Coatings in development and supporting of several paint assortments. He built up a department application research with his knowledge about behaviour psychology (mature education), in which he rounded off his degree at the University of Amsterdam in 1981. Since 1991 he was Technical Secretary of the Dutch Paint and Printing Ink Association. In that time he introduced Coating Care in The Netherlands and assisted on many regulations in The Netherlands and in the European Community, for example the Product Directive 2004/42/EC. At this moment he is freelance editor, teacher in the Elesevier/Reed organization in Coating Technology and has his own Coating Technology Consultancy office in The Netherlands.

Adrie Winkelaar: Coatings Basics
© Copyright 2009 by Vincentz Network, Hannover, Germany
ISBN 978-3-86630-851-0

Index

Adrie Winkelaar: Coatings Basics
© Copyright 2009 by Vincentz Network, Hannover, Germany
ISBN 978-3-86630-851-0